COPYRIGHT © 2002 Thomson Learning, Inc.

Thomson Learning™ is a trademark used herein under license.

ALL RIGHTS RESERVED. No part of this work covered by the copyright hereon may be reproduced or used in any form or by any means—graphic, electronic, or mechanical, including, but not limited to, photocopying, recording, taping, Web distribution, information networks, or information storage and retrieval systems—without the written permission of the publisher.

Printed in Canada

2 3 4 5 6 7 05 04 03 02

ISBN: 0-03-033716-X

> **For more information about our products, contact us at:**
> **Thomson Learning Academic Resource Center**
> **1-800-423-0563**
>
> **For permission to use material from this text, contact us by:**
> **Phone: 1-800-730-2214**
> **Fax: 1-800-731-2215**
> **Web: www.thomsonrights.com**

Asia
Thomson Learning
60 Albert Complex, #15-01
Alpert Complex
Singapore 189969

Australia
Nelson Thomson Learning
102 Dodds Street
South Street
South Melbourne, Victoria 3205
Australia

Canada
Nelson Thomson Learning
1120 Birchmount Road
Toronto, Ontario M1K 5G4
Canada

Europe/Middle East/South Africa
Thomson Learning
Berkshire House
168-173 High Holborn
London WC1 V7AA
United Kingdom

Latin America
Thomson Learning
Seneca, 53
Colonia Polanco
11560 Mexico D.F.
Mexico

Spain
Paraninfo Thomson Learning
Calle/Magallanes, 25
28015 Madrid, Spain

Core Concepts in College Physics Workbook

CONTENTS

Introduction vii

USER'S GUIDE

System Requirements 1
Macintosh 1
Windows 1

**Instructions for
College Physics CD Installation** 2
Macintosh 2
Windows 2
   Windows 95 and NT 4.0 Systems 2
   Windows 3.1 or NT 3.5 Systems 3

Release Notes for Version 1.0 4

Using the CD-ROM 5

PROBLEMS

MODULE 1 **Problem Solving in Physics** 11

MODULE 2 **Vectors** 12
Problem 1 (Screen 2.2) Coordinate Systems 13
Problem 2 (Screen 2.4) Vector Addition and Subtraction 14
Problem 3 (Screen 2.5) Vector Components and Unit Vectors 15
Problem 4 (Screen 2.5) Vector Components and Unit Vectors 16
Problem 5 (Screen 2.6) The Dot Product 18
Problem 6 (Screen 2.6) The Dot Product 19

MODULE 3 **Kinematics** 20
Problem 7 (Screen 3.2) Displacement, Velocity, and Speed 22
Problem 8 (Screen 3.2) Displacement, Velocity, and Speed 24
Problem 9 (Screen 3.4) One-Dimensional Motion at Constant Acceleration 26
Problem 10 (Screen 3.4) One-Dimensional Motion at Constant Acceleration 28
Problem 11 (Screen 3.5) Projectile Motion 30
Problem 12 (Screen 3.5) Projectile Motion 32
Problem 13 (Screen 3.6) Uniform Circular Motion 34
Problem 14 (Screen 3.7) Relative Motion 36

MODULE 4 Forces 38

Problem 15 (Screen 4.2) Motion, Newton's First Law, and Force 40
Problem 16 (Screen 4.4) Newton's Second Law 41
Problem 17 (Screen 4.5) Newton's Third Law 42
Problem 18 (Screen 4.5) Newton's Third Law 44
Problem 19 (Screen 4.6) Newton's Second Law 45
Problem 20 (Screen 4.6) Free-Body Diagrams 47
Problem 21 (Screen 4.6) Free-Body Diagrams 48
Problem 22 (Screen 4.6) Free-Body Diagrams 49
Problem 23 (Screen 4.7) Centripetal Force 51
Problem 24 (Screen 4.8) Fictitious Forces: Motion in Accelerated Reference Frames 52

MODULE 5 Work and Energy 54

Problem 25 (Screen 5.2) Work 56
Problem 26 (Screen 5.3) Important Examples of Work: Gravity and Springs 58
Problem 27 (Screen 5.5) Energy 59
Problem 28 (Screen 5.6) Conservative Forces 60
Problem 29 (Screen 5.7) The Work-Energy Theorem 62
Problem 30 (Screen 5.7) The Work-Energy Theorem 65
Problem 31 (Screen 5.8) Power 67
Problem 32 (Screen 5.8) Power 69
Problem 33 (Screen 5.9) Conservation of Energy 71

MODULE 6 Linear Momentum 73

Problem 34 (Screen 6.2) Conservation of Momentum 75
Problem 35 (Screen 6.4) Impulse 77
Problem 36 (Screen 6.5) Perfectly Inelastic Collisions 79
Problem 37 (Screen 6.5) Perfectly Inelastic Collisions 81
Problem 38 (Screen 6.6) Perfectly Elastic Collisions 83
Problem 39 (Screen 6.6) Perfectly Elastic Collisions 85
Problem 40 (Screen 6.7) Center of Mass 87

MODULE 7 Rotational Mechanics 88

Problem 41 (Screen 7.2) Rotational Kinematics 90
Problem 42 (Screen 7.2) Rotational Kinematics 92
Problem 43 (Screen 7.3) Rotational Kinetic Energy 94
Problem 44 (Screen 7.5) Torque 96
Problem 45 (Screen 7.6) Work and Energy in Rotational Motion 98
Problem 46 (Screen 7.7) Rolling Motion 100
Problem 47 (Screen 7.9) Conservation of Angular Momentum 102

MODULE 8 Simple Harmonic Motion and Waves 104

Problem 48 (Screen 8.3) Simple Harmonic Motion 106
Problem 49 (Screen 8.6) Physical Nature of Waves 108
Problem 50 (Screen 8.6) Frequency, Wavelength, and Wave Speed 109
Problem 51 (Screen 8.6) Frequency, Wavelength, and Wave Speed 110
Problem 52 (Screen 8.8) Mathematical Nature of Waves 111
Problem 53 (Screen 8.9) Mathematical Nature of Waves 113
Problem 54 (Screen 8.10) Hooke's Law and the Equation of Motion 114
Problem 55 (Screen 8.11) SHM and Waves in the Real World 117
Problem 56 (Screen 8.14) Wave Speed 118

Core Concepts in College Physics Workbook

MODULE 9 **Wave Behavior** *120*

Problem 57 (Screen 9.2) Speed of a Wave in a Medium *122*
Problem 58 (Screen 9.5) Energy and Power in Waves *124*
Problem 59 (Screen 9.7) Superposition and Interference *125*
Problem 60 (Screen 9.8) Standing Waves *126*
Problem 61 (Screen 9.9) Standing Waves—Wave Fixed at Both Ends *128*
Problem 62 (Screen 9.9) Standing Waves—Wave with One Fixed End and One Free End *130*

MODULE 10 **Thermodynamics** *132*

Problem 63 (Screen 10.3) Basic Concepts of Thermodynamics *135*
Problem 64 (Screen 10.5) The Ideal Gas *137*
Problem 65 (Screen 10.6) The First Law of Thermodynamics *138*
Problem 66 (Screen 10.9) Carnot Engines *140*
Problem 67 (Screen 10.9) Carnot Engines—The Heat Pump *142*
Problem 68 (Screen 10.10) Entropy *143*

MODULE 11 **The Electric Field** *145*

Problem 69 (Screen 11.4) Coulomb's Law *148*
Problem 70 (Screen 11.5) The Electric Field and Field Lines *150*
Problem 71 (Screen 11.6) Gauss's Law *152*
Problem 72 (Screen 11.7) Examples of the Electric Field *153*
Problem 73 (Screen 11.8) Electric Potential *154*
Problem 74 (Screen 11.9) Electric Field and Electric Potential *156*

MODULE 12 **The Magnetic Field** *158*

Problem 75 (Screen 12.3) Magnetic Force on a Moving Charge *161*
Problem 76 (Screen 12.3) The Lorentz Force *163*
Problem 77 (Screen 12.4) Ampère's Law *165*
Problem 78 (Screen 12.5) Magnetic Flux and Gauss's Law for Magnetism *167*
Problem 79 (Screen 12.7) Faraday's Law of Induction and Lenz's Law *169*
Problem 80 (Screen 12.7) Faraday's Law *170*

MODULE 13 **Electric Circuits** *172*

Problem 81 (Screen 13.3) Voltage, Resistance, and Ohm's Law *175*
Problem 82 (Screen 13.4) Circuit Analysis and Kirchhoff's Laws *177*
Problem 83 (Screen 13.4) Circuit Analysis and Kirchhoff's Laws *179*
Problem 84 (Screen 13.5) Capacitors *180*
Problem 85 (Screen 13.6) Inductors *182*
Problem 86 (Screen 13.7) Circuits Containing Resistors, Inductors, and Capacitors *183*
Problem 87 (Screen 13.7) Circuits Containing Resistors, Inductors, and Capacitors *185*

MODULE 14 **Geometric Optics** *186*

Problem 88 (Screen 14.3) Reflection *188*
Problem 89 (Screen 14.4) Snell's Law *189*
Problem 90 (Screen 14.4) Snell's Law *190*
Problem 91 (Screen 14.5) Total Internal Reflection *191*
Problem 92 (Screen 14.7) Flat and Spherical Mirrors *192*
Problem 93 (Screen 14.8) Thin Lenses *193*

SOLUTIONS

MODULE 2 **Vectors** 195
[Problems 1-6]

MODULE 3 **Kinematics** 198
[Problems 7-14]

MODULE 4 **Forces** 205
[Problems 15-24]

MODULE 5 **Work and Energy** 212
[Problems 25-33]

MODULE 6 **Linear Momentum** 222
[Problems 34-40]

MODULE 7 **Rotational Mechanics** 228
[Problems 41-47]

MODULE 8 **Simple Harmonic Motion and Waves** 235
[Problems 48-55]

MODULE 9 **Wave Behavior** 242
[Problems 57-62]

MODULE 10 **Thermodynamics** 245
[Problems 63-68]

MODULE 11 **The Electric Field** 250
[Problems 69-74]

MODULE 12 **The Magnetic Field** 255
[Problems 75-80]

MODULE 13 **Electric Circuits** 261
[Problems 81-87]

MODULE 14 **Geometric Optics** 266
[Problems 88-93]

APPENDIX

Reference Tables

Some Fundamental Constants 271
Solar System Data 273
Physical Data Often Used 273
Some Prefixes for Powers of Ten 274
Standard Abbreviations and Symbols of Units 274
Mathematical Symbols Used in the Text and
    Their Meaning 275

Useful Conversions 276
The Greek Alphabet 276
Conversion Factors 277
Symbols, Dimensions, and Units of
    Physical Quantities 279
SI Base Units 282
Some Derived SI Units 282

Core Concepts in College Physics Workbook

INTRODUCTION

Core Concepts in College Physics for Macintosh® and Windows™ is an interactive, three-disc CD-ROM presentation of introductory, algebra and trigonometry-based physics for college and university students.

FEATURES OF THE CD-ROM

The presentation:

- Uses live video, animation, interactive graphics, audio, and text to teach fundamental principles of introductory physics.
- Applies the presented concepts to real world phenomena.
- Bridges physical principles to the mathematics that describe them.
- Provides tools for learning and doing physics.

Our goal is to help you develop a deep and practical understanding of physical phenomena, to directly assist you in your study of physics. Because problem-solving is such an essential skill for success in physics, we have also included worked problems and "pop questions" within the presentation.

This three-disc set contains 14 modules (similar to chapters in a textbook) and is accompanied by this workbook, which can be used in addition to any other general physics text.

Disc 1 contains Modules 1-5.
Disc 2 contains Modules 6-9.
Disc 3 contains Modules 10-14.

These discs also include:

- Tools such as a Unit Converter and Physical Constants table
- Active indexes of contents and equations
- Worked Sample Problems and Pop Questions

INTERACTIVE PRESENTATION Information is presented as "screens" within modules. Each screen in a module introduces a key concept or a set of related concepts. Complete instructions for using this presentation are in the section of this workbook entitled *Using the CD-ROM*, and in the instructions that accompany the discs.

WORKBOOK The workbook is organized around the main concept screens, each of which is numbered by the module and screen. The questions in the workbook can be answered after reading the text on the screen (and related screens), viewing the media, and working through associated problems.

USER'S GUIDE

SYSTEM REQUIREMENTS

To use the applications included on the CD, your system must meet the following requirements:

Macintosh® operation requires:

- Power Macintosh running System 7.5.5 or above
- 800 × 600 pixel resolution; thousands of colors
- QuickTime 4 (Provided on CD-ROM)
  It is important that the full version of QuickTime be installed.

Windows™ operation requires:

- Pentium class processor running Windows 95, 98, NT, ME or 2000
- 800 × 600 pixel resolution; 16-bit color
- Windows-compatible sound card
- QuickTime 4 (Provided on CD-ROM)
  It is important that the full version of QuickTime be installed.

INSTRUCTIONS FOR PHYSICS CD INSTALLATION

You do not need to install the *Core Concepts in College Physics* program as long as your computer has QuickTime 4 or higher. If your computer does not have QuickTime 4 or higher, you can install it from any of the three discs by double clicking on its icon, which is in the Tools directory of the CD-ROM.

QuickTime Installation Procedure for Macintosh

If you do not have QuickTime 4 or higher, insert any of the three discs and open the Tools folder.

Double-click on the QuickTime 4 icon.

Double-click on the QuickTime Installer found on the desktop.

To complete installation, double-click on Install QuickTime.

Windows Installation Issues

In order for this program to function correctly on a given machine, the machine must have the two dynamic link library files msvcrt.dll and msvfw32.dll installed in the system directory. The resource file msvcrt.dll is included by default with all 32-bit Microsoft Windows™ operating system installations with the exception of Windows 95A. This first version of the Windows 95 operating system did not include this file.

The resource file msvfw32.dll is included by default with all Microsoft Windows™ 98 and Windows NT operating system installations. When installing any version of Windows 95, the user has the option of performing a custom installation. If custom installation is chosen, the user can choose to install the system multimedia components.

Core Concepts in College Physics Workbook

If the option is selected NOT to install the system multimedia components, msvfw32.dll will not be installed on that machine. Therefore, some machines running Windows 95 may be missing the library file.

Solution
Unzip D7projector_dlls_installer.exe located in the Tools directory of the CD-ROM (any disc). Install the unzipped files into the machine's system directory, if the files are not already present. For Windows NT machines, the system directory is typically located at "C:\WinNT\System32." On Windows 95 and 98 machines, the system directory is typically located at "C:\Windows\System."

Windows 2000 Issue
When using the program for long periods of time in Windows 2000, the cursor may begin to "flutter." If this occurs, restart the program (or switch to another program and back again) to eliminate this problem.

CORE CONCEPTS IN
PHYSICS CD-ROM

The *Core Concepts in College Physics* CD-ROM is a complete multimedia presentation of college-level calculus-based introductory physics.

Starting the Presentation for Macintosh
Once the desired disc is in the CD-ROM drive, open the CD-ROM by double-clicking on its icon.

To begin the presentation, double-click on the *Physics* icon.

Starting the Presentation for Windows 95, 98, NT, ME or 2000 Systems
Once the desired disc is in the CD-ROM drive, open "My Computer" and double-click on the CD-ROM drive icon.

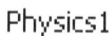

To begin the presentation, double-click on the *Physics* icon.

Using the Presentation
A title screen appears with a "Production Credits" bar at the screen's lower right corner. To view credits, move hand cursor and click on the bar. To view the *Contents* screen, click anywhere else on the title screen or credits screen.

The mouse is used for all navigation. Navigation within the presentation is accomplished by a single click of the mouse.

A pointing finger cursor indicates an active area.

Inactive screen areas are indicated by the arrow cursor.

[Title Screen]

Core Concepts in College Physics Workbook

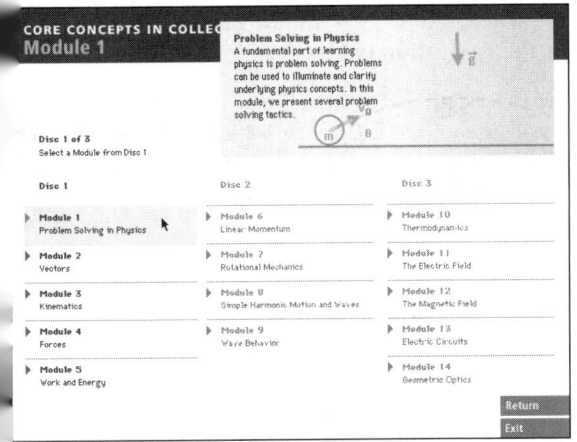

[Main Contents Screen]

The presentation is divided into modules, with Modules 1–5 on Disc 1, Modules 6–9 on Disc 2, and Modules 10–14 on Disc 3.

Click on the desired module to launch it from the *Contents* screen.

In the Contents, all active material is highlighted; all inactive material is dimmed. The material from discs other than the one in use can be accessed only by inserting the desired disc, although synopses for all modules are viewable on each disc by rolling the mouse over the module title.

The first screen of every module is a *Contents and Introduction* screen. From this screen, view the introductory "touchstone" movie or click on a topic name to open the desired Main Screen.

Each module is organized into a series of Main Screens, which address a single topic or a group of closely related topics. The *Contents and Introduction* screen provides a list of that module's Main Screens, as well as the introductory movie.

[Contents and Introduction Screen]

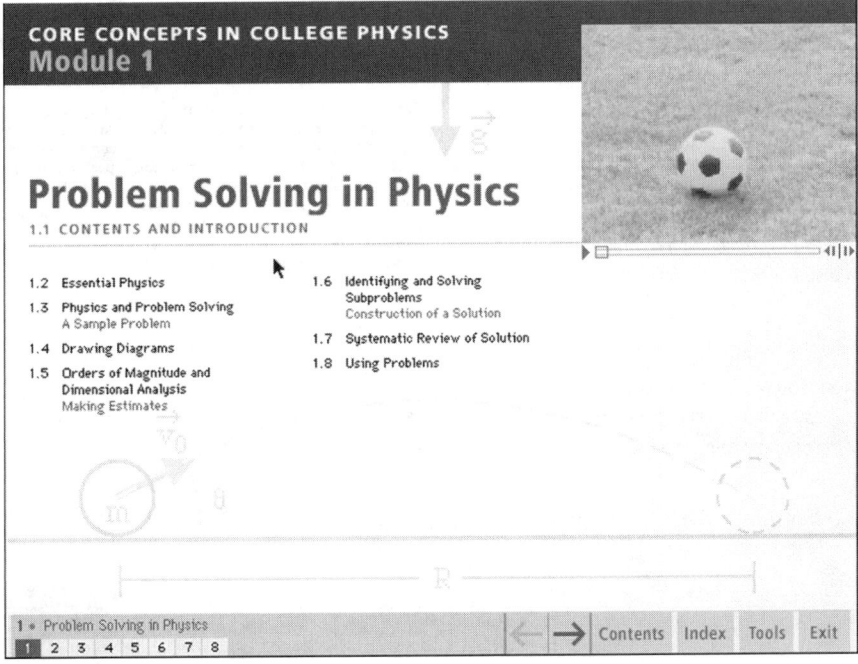

The Core Concepts in College Physics CD-ROM

Navigation Bar Operation

At the bottom of the screen is the navigation bar.

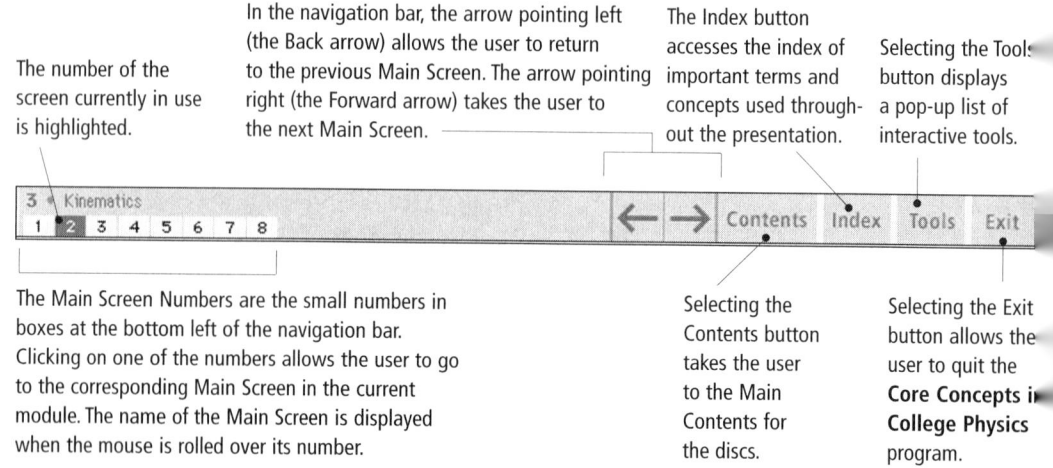

The number of the screen currently in use is highlighted.

In the navigation bar, the arrow pointing left (the Back arrow) allows the user to return to the previous Main Screen. The arrow pointing right (the Forward arrow) takes the user to the next Main Screen.

The Index button accesses the index of important terms and concepts used throughout the presentation.

Selecting the Tools button displays a pop-up list of interactive tools.

The Main Screen Numbers are the small numbers in boxes at the bottom left of the navigation bar. Clicking on one of the numbers allows the user to go to the corresponding Main Screen in the current module. The name of the Main Screen is displayed when the mouse is rolled over its number.

Selecting the Contents button takes the user to the Main Contents for the discs.

Selecting the Exit button allows the user to quit the **Core Concepts in College Physics** program.

Main Screens

Main Screens are accessed either from the Table of Contents on the Module *Contents and Introduction* screen or from the Navigation Bar at the bottom of each screen in a Module. Each Main Screen includes several features such as video or audio clips, problems, sidebars, tables, math-in-detail banners, or animated simulations that provide information about the current topic.

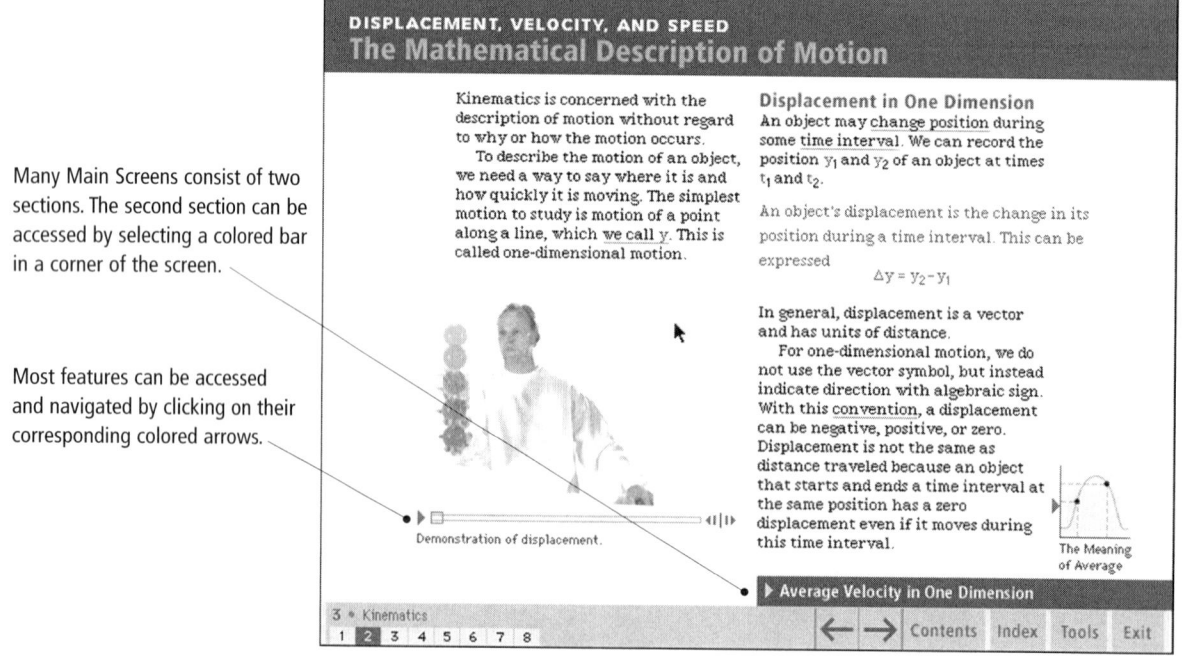

Many Main Screens consist of two sections. The second section can be accessed by selecting a colored bar in a corner of the screen.

Most features can be accessed and navigated by clicking on their corresponding colored arrows.

Core Concepts in College Physics Workbook

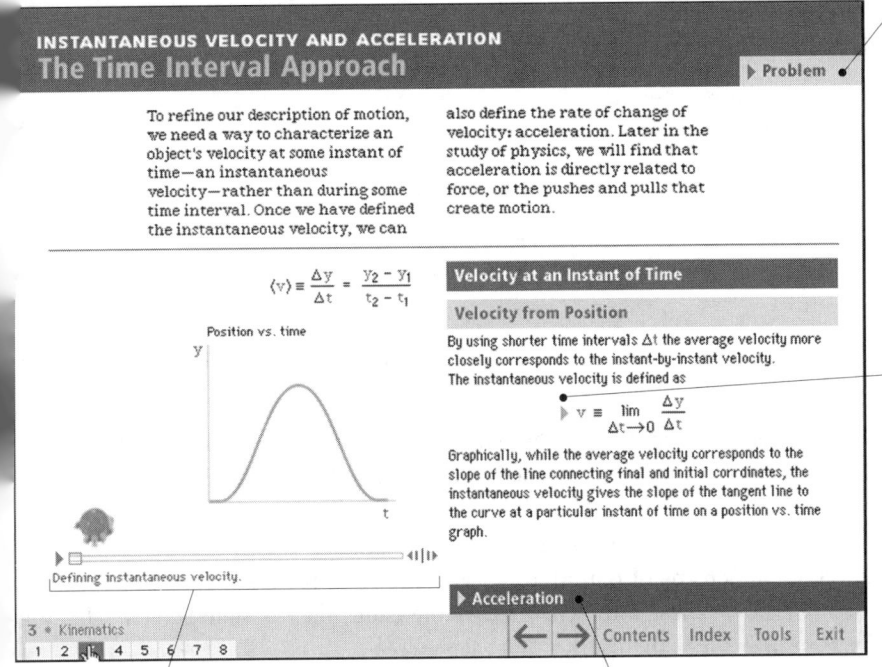

The Problem button is found in the top right corner of some screens. Select it to access a problem related to the current topic.

The arrows in the screen are used to access more information or to initiate an action such as playing an animation, or displaying mathematics-in-detail banners.

At the lower edge of most movies is a sliding control bar with play/pause button, which allows the user to see the movie with narration. The slider can be manipulated using the mouse to move forward or backward through the clip.

A Section Bar is often found at the bottom right of Main Screens; selecting it accesses the next section of the Main Screen.

Pop-up Questions are indicated by the Q icon on many screens. Select the icon for the question. To view answer choices, select the "A" in the question box. The correct answer appears after you select your answer choice.

Place the cursor over any underlined text to access a definition or explanation of that term.

[Tools button]

Accessing the Tools Menu
Interactive tools are available from every Presentation Screen, accessed from the Tools button in the Navigation Bar. Selecting and holding the button displays a pop-up list of the tools: the Unit Converter, Equations index, Physical Constants and Physical Data tables, and more. To access a tool, roll over its name and release.

To return to the presentation screen from the tool, click on the return bar at the bottom right of the menu.

The Core Concepts in College Physics CD-ROM

Reference Section

This section is organized by:

- Navigation Functions
- Media Access Functions
- Index

Navigation Functions

Forward and reverse arrows allow the user to move to the next or previous Main Screen.

[Forward/Backward Arrows]

Note: The Module Opener contains no active reverse arrow, and no active forward arrow appears on the last screen of a Module.

The Contents button allows the user to return to the *Contents* screen. It also allows the user to move from one Module to another on the active disc.

[Contents button]

At the bottom of every Module screen is a sequence of numbers corresponding to the Main Screens of that Module. The current screen is highlighted. Moving the mouse over these numbers displays the name of each Main Screen. Clicking on a number takes the user to that Main Screen.

[Main Screen Numbers]

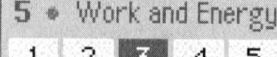

The Exit button quits the presentation.

[Exit button]

Media Access Functions

Small colored arrows, found throughout the *Core Concepts in Physics* presentation, initiate some action such as accessing mathematics-in-detail banners, sidebars, or other features.

[Colored Arrows]

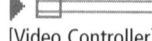
[Video Controller]

Video buttons allow the user to play, stop, pause, and replay video or animation clips. Most clips have a sliding controller that allows users to advance or reverse the clip.

[Pause]

Note: If the video plays poorly or drops frames, or if the audio tends to "cut-out," make sure you have quit any open applications to free up additional RAM. If the problem persists, try running the disc on a computer with more memory or a faster CD-ROM drive.

Index

The Navigation Bar also includes an Index button.

The Index of important terms and concepts allows the user to find any important topic or term featured within the presentation. Select the colored numbers to move to a reference on the current disc.

[Index button]

[Index Screen]

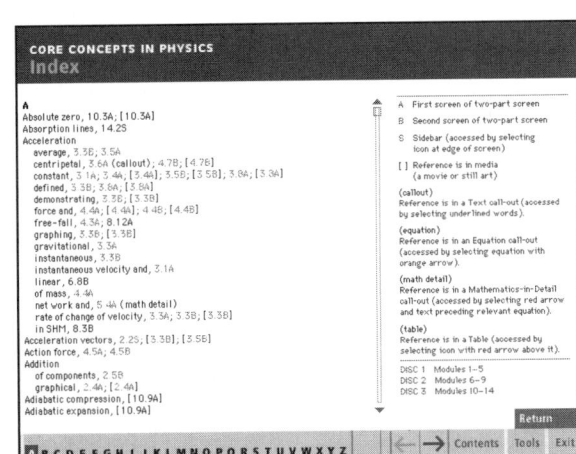

TROUBLESHOOTING AND TECHNICAL SUPPORT

For technical support assistance, contact us by:

1. Telephone support.   1-800-423-0563
                         Monday-Friday, 8:30 AM to 6:00 PM Eastern time

2. FAX.   1-859-647-5045
You may write out your question and FAX it to the number above.

3. E-mail.   support@kdc.com

For updates on software and other titles, visit
http://www.thomsonlearning.com.

MODULE 1

Problem Solving in Physics

It has been said that practice is the best instructor. In the study of physics, "practice" means solving problems. This workbook is designed to help you get practice.

Before tackling any physics problem, you will find it useful to equip yourself with a set of mental tools. We strongly recommend that you view the *Problem Solving in Physics* module as a first step in acquiring these tools, and to continue to refer back to it as you progress in your studies.

Here is a checklist of reminders for solving most of the problems you are likely to encounter in your physics course:

- Draw a diagram
- Estimate order of magnitude
- Perform dimensional analysis
- Identify and solve subproblems
- Construct a final answer
- Systematically review your answer
  - Check order of magnitude
  - Check units/dimensions
  - Interpret your equations for "sanity"
  - Check special cases

You will very likely wish to adapt and build on these tools as you develop your own strategies and tactics for solving problems. A key point to remember is that learning physics does not demand memorizing innumerable equations; instead, it calls for the building of bridges between various concepts and principles in physics.

MODULE 2 — Vectors

INTRODUCTION

Physical phenomena can often be quantized by magnitude alone (e.g., the temperature at a given point in space). Other times, they can be specified by both a direction and a magnitude (e.g., the wind's speed and direction at the same point). Quantities that have both magnitude and direction are described by vectors. In this module, we introduce vectors and some of their applications and operations (such as addition, subtraction, and multiplication). We also discuss polar and cartesian coordinate systems, and why choosing a particular coordinate system can simplify an analysis.

DEFINITIONS

We use the three trigonometric functions:

$$\sin \theta = \frac{\text{opposite}}{\text{hypotenuse}} \qquad \cos \theta = \frac{\text{adjacent}}{\text{hypotenuse}} \qquad \tan \theta = \frac{\text{opposite}}{\text{adjacent}}$$

and the Pythagorean theorem:

$$(\text{hypotenuse})^2 = (\text{adjacent})^2 + (\text{opposite})^2$$

Vector components.

If $\vec{A} + \vec{B} = \vec{C}$, then $A_x + B_x = C_x$ and $A_y + B_y = C_y$

Scalar (dot) product.

$\vec{A} \cdot \vec{B} = |\vec{A}||\vec{B}| \cos \theta$, where $\theta =$ angle between $\vec{A}$ and $\vec{B}$

$\vec{A} \cdot \vec{B} = A_x B_x + A_y B_y + A_z B_z$

Core Concepts in College Physics Workbook

PROBLEM 1 (SCREEN 2.2)

Coordinate Systems

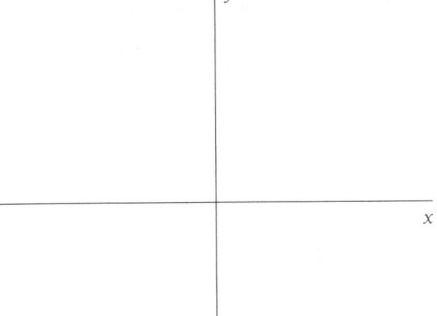

Problem Description
Two points in a plane have polar coordinates $(r, \theta) = (2.50$ m, $30.0°)$ and $(3.80$ m, $120.0°)$, respectively. Determine the cartesian coordinates of these points and the distance between them.

Before we begin...

1. Draw a diagram indicating the two vectors as $\vec{A}$ and $\vec{B}$.

2. How are the cartesian coordinates x and y related to the polar coordinates, r and θ?

Solving the problem

3. For each vector, $\vec{A}$ and $\vec{B}$, find the x and y components.

4. The vector separating the two points is $\vec{B} - \vec{A}$. Find the x and y coordinates of $\vec{B} - \vec{A}$.

5. Use the Pythagorean theorem to find the distance, which is the magnitude of $\vec{B} - \vec{A}$.

Module 2 **Vectors**

PROBLEM 2 (SCREEN 2.4)

Vector Addition and Subtraction

Problem Description

A force $\vec{F_1}$ of magnitude 6.00 units acts on an object at the origin in a direction 30.0° above the positive x axis. A second force $\vec{F_2}$ of magnitude 5.00 units acts on the same object in the direction of the positive y axis. Use a graph to determine the magnitude and direction of the resultant force $\vec{F_1} + \vec{F_2}$.

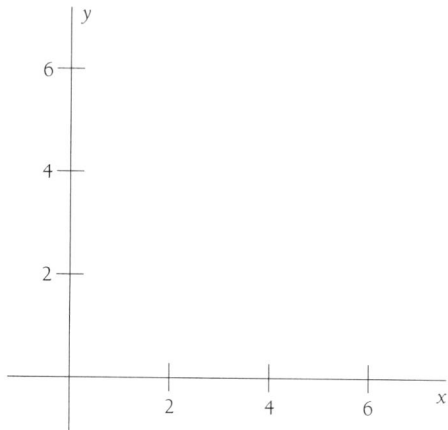

Before we begin...   1. Draw the forces from the origin in the diagram shown here.

Solving the problem   2. Given the forces as drawn, use graphical addition to find the resultant force.

Core Concepts in College Physics Workbook

PROBLEM 3   (SCREEN 2.5)

Vector Components

Problem Description

A displacement vector lying in the *xy* plane has a magnitude of 50.0 m and is directed at an angle of 120.0° above the positive *x* axis. What are the rectangular components of this vector?

Before we begin...

1. The vector can be expressed in polar coordinate form as $\vec{A}$ = (50.0 m, 120.0°). How are the rectangular coordinates of a vector related to the polar coordinates?

2. Draw the vector on the above axis system. Drop perpendicular lines from the tip of the vector to the *x* axis and to the *y* axis. These projections give the length of A_x and A_y, respectively.

Solving the problem

3. Use the information in the above answer to find the rectangular coordinates A_x and A_y.

PROBLEM 4 (SCREEN 2.5)

Vector Components

Problem Description

Instructions for finding a buried treasure include the following: Go 75 paces at 240°, turn to 135° and walk 125 paces, then travel 100 paces at 160°. Determine the resultant displacement from the starting point.

Before we begin...

1. Draw a sketch of the problem, using the graphical method of adding vectors.

2. Express the three displacement vectors $\vec{A}$, $\vec{B}$, and $\vec{C}$ in polar coordinates.

   $\vec{A} =$ $\vec{B} =$ $\vec{C} =$

Solving the problem

3. Find the x and y components of the three vectors.

Core Concepts in College Physics Workbook

4. Add the x components together to get the x component of the total displacement. Do the same for the y components.

5. Use the Pythagorean theorem to find the magnitude of the resultant vector.

6. Use a suitable trigonometric function to determine the angle that the resultant vector makes with respect to the x axis.

PROBLEM 5 (SCREEN 2.6)

The Dot Product

Problem Description
Vector $\vec{A}$ extends from the origin to a point having polar coordinates (7, 70°) and vector $\vec{B}$ extends from the origin to a point having polar coordinates (4, 130°). Find $\vec{A} \cdot \vec{B}$.

Before we begin...

1. Sketch the vectors $\vec{A}$ and $\vec{B}$ on the coordinate system above. Indicate the angle that is formed between the two vectors.

2. What are the magnitudes of the two vectors?

   $|\vec{A}| = $ $|\vec{B}| = $

3. Is $\vec{A} \cdot \vec{B}$ a vector or a scalar quantity?

Solving the problem

$\vec{A} \cdot \vec{B}$ is the scalar product of the two vectors. In this problem, the vectors are expressed in polar coordinates. The scalar product is calculated according to the relation:

$$\vec{A} \cdot \vec{B} = |\vec{A}||\vec{B}| \cos \theta$$

where θ is the angle between $\vec{A}$ and $\vec{B}$.

4. Use the information gathered above to compute the value.

PROBLEM 6 (SCREEN 2.6)

The Dot Product

Problem Description
Vector $\vec{A}$ is 2.0 units long and points in the positive y direction. Vector $\vec{B}$ has a negative x component 5.0 units long, a positive y component 3.0 units long, and no z component. Find $\vec{A} \cdot \vec{B}$ and the angle between the vectors.

Before we begin...

1. In this problem, the vectors are expressed in a different form than in the previous example. What form does this problem use?

2. Identify the x, y, and z components of each vector:

   $A_x =$ $A_y =$ $A_z =$

   $B_x =$ $B_y =$ $B_z =$

Solving the problem

3. The scalar product in rectilinear coordinates (x, y, z) is computed according to the relationship:

   $$\vec{A} \cdot \vec{B} = A_x B_x + A_y B_y + A_z B_z$$

   Compute the scalar product.

   Notice that the scalar product is a quantity without direction. It is not a vector. If you know the scalar product, you can use the relationship:

   $$\vec{A} \cdot \vec{B} = |\vec{A}||\vec{B}| \cos \theta$$

   to find θ, the angle between the two. In order to find θ, what do you need to compute before substituting into the relationship?

4. Calculate the angle θ.

MODULE 3 Kinematics

INTRODUCTION

In this module, we describe and quantify the motion of objects in one- and two-dimensions. This is a necessary first step before we can hope to learn about the causes of motion, which we begin to explore in Module 4, *Forces*.

The concepts used to describe motion, such as displacement, velocity, and acceleration, are vectors. For motion in one direction (i.e., back and forth along a straight-line path), a simple + or − sign indicates the direction of these vectors. For two- and three-dimensional motion, we use the full set of vector operations we studied in Module 2, *Vectors*.

DEFINITIONS

Displacement. The change in position of an object represented by Δx in one-dimensional motion and $\Delta \vec{r}$ in more than one dimension. It is a vector quantity that is computed by

$$\Delta \vec{r} = \vec{r}_f - \vec{r}_i$$

Velocity, v. The rate of change of displacement with respect to time. The average velocity during a time interval Δt is defined as

$$\langle \vec{v} \rangle = \frac{\Delta \vec{r}}{\Delta t}$$

The instantaneous velocity is defined as

$$\vec{v} \equiv \lim_{\Delta t \to 0} \frac{\Delta \vec{r}}{\Delta t}$$

On a graph of displacement as a function of time, the slope of the line connecting two points gives the average velocity, while the slope of the tangent line to the displacement at the time gives the instantaneous velocity.

Acceleration, $\vec{a}$. The rate of change of velocity with respect to time. The average acceleration is defined as

$$\langle \vec{a} \rangle = \frac{\Delta \vec{v}}{\Delta t}$$

and the instantaneous acceleration is defined as

$$\vec{a} \equiv \lim_{\Delta t \to 0} \frac{\Delta \vec{v}}{\Delta t}$$

USEFUL EQUATIONS
(*Note:* In one dimension we use x or y rather than r for the position coordinate.)

The position $\vec{r}$ can be expressed in terms of its components along the x and y axes.

$$\vec{r} = \vec{r}_x + \vec{r}_y = (x, y)$$

For constant acceleration

$$\vec{v} = \vec{v}_0 + \vec{a}t$$

$$\vec{r} = \vec{r}_0 + \vec{v}_0 t + \frac{1}{2}\vec{a}t^2$$

$$\vec{v}^2 = \vec{v}_0^2 + 2\vec{a} \cdot \vec{r}$$

For uniform circular motion, $\vec{a}_c$, the centripetal acceleration, is directed toward the center of the circle, with magnitude

$$a_c = \frac{v^2}{r}$$

For nonuniform circular motion

$$\vec{a} = \vec{a}_t + \vec{a}_c$$

where $\vec{a}_t$ is the tangential component and $\vec{a}_c$ the centripetal, or radial, component of the acceleration.

Module 3   Kinematics

PROBLEM 7 (SCREEN 3.2)

Displacement, Velocity, and Speed

Problem Description

The velocity of a particle as a function of time is shown. At $t = 0$, the particle is at $x = 0$. Sketch the acceleration as a function of time. Determine the average acceleration of the particle from time $t = 2.0$ s to $t = 8.0$ s. Determine the instantaneous acceleration of the particle at $t = 4.0$ s.

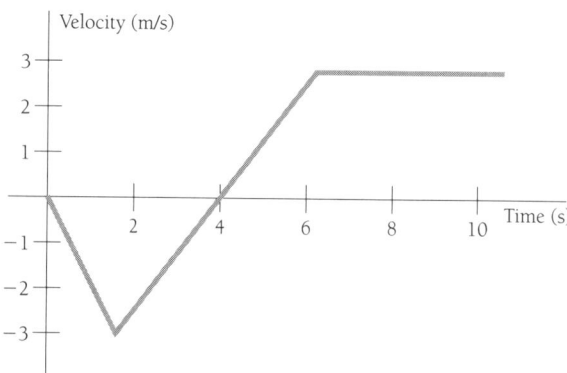

Before we begin...

The graph of velocity as a function of time given in this problem consists of three straight line segments, from (0 s, 0 m/s) to (2 s, −3 m/s) to (6 s, 3 m/s) to (8 s, 3 m/s).

1. How is the acceleration vector related to velocity vs. time?

Solving the problem

2. Does the acceleration change in the time between $t = 0$ and $t = 2$ s? What is the value of the acceleration?

3. Does the acceleration change in the time between $t = 2$ and $t = 6$ s? What is the value of the acceleration?

Core Concepts in College Physics Workbook

4. Does the acceleration change in the time between $t = 6$ and $t = 8$ s?

   What is the value of the acceleration?

5. You can now graph acceleration vs. time, using the answers to the above questions.

Because the acceleration is equal to the rate of change of the velocity, we can use the relation

$$\langle \vec{a} \rangle = \frac{\Delta \vec{v}}{\Delta t}$$

to find the average acceleration over any time interval.

6. Use the data from the original graph to determine the particle's average acceleration from time $t = 2.0$ s to $t = 8.0$ s.

7. Use the slope of the tangent line to the velocity vs. time graph to evaluate the acceleration at the particular point in time $t = 4.0$ s.

PROBLEM 8   (SCREEN 3.2)

Displacement, Velocity, and Speed

Problem Description

An athlete swims the length of a 50.0-meter-long pool in 20.0 s and makes the return trip in 22.0 seconds. Determine her average velocities for the first half of the swim, for the second half, and for the round trip.

Before we begin...   Decide what coordinate axes you will use to describe the swimmer's motion. We will take the direction she swims during the first leg of her trip to be the $+x$ direction.

1. Identify the given information (*Hint*: Displacement can be positive or negative, depending on its direction):

    displacement during first leg of trip $\Delta x_1 =$
    elapsed time during first leg of trip $\Delta t_1 =$

    displacement during first leg of trip $\Delta x_2 =$
    elapsed time during first leg of trip $\Delta t_2 =$

2. For which leg of the trip do you expect to find a faster velocity? (Later, you can use this prediction as a quick check that your answer is working out well.)

Solving the problem

3. What is the equation relating average velocity to the displacement and elapsed time?

4. Use this equation to find the average velocities $\langle v_1 \rangle$ and $\langle v_2 \rangle$ during the first and second legs of the swim, respectively. (Do they satisfy your prediction from step 2?)

5. To find the average velocity during the entire round trip, we need to identify two or more quantities:

   displacement during entire round trip $\Delta x_{tot} =$
   elapsed time during entire round trip $\Delta t_{tot} =$

6. What was the swimmer's average velocity during the entire swim? How do you interpret this answer?

PROBLEM 9 (SCREEN 3.4)

One-Dimensional Motion at Constant Acceleration

Problem Description

Two bicyclists, Bicyclist A and Bicyclist B, are traveling along a straight road. Bicyclist A has stopped to wait for Bicyclist B, who is traveling at a constant speed of 8 m/s. Bicyclist A remains at rest until the instant when Bicyclist B passes her; she then begins accelerating at a constant rate of 2 m/s². How long does it take Bicyclist A to catch up to Bicyclist B? How fast is she traveling at the moment she passes him?

Before we begin...

Decide what coordinate axes you will use to describe the cyclists' motion. We will take the position where Bicyclist A begins moving to be the origin, and the direction that they move to be the $+x$ direction.

1. Identify the given information (including the units of each quantity):

	Bicyclist A	Bicyclist B
initial position	$x_{0A} =$	$x_{0B} =$
initial velocity	$v_{0A} =$	$v_{0B} =$
acceleration	$a_A =$	$a_B =$

2. Try to estimate the form of the answer by dimensional analysis. The two relevant quantities are Bicyclist A's acceleration (in m/s²) and Bicyclist B's constant velocity (in m/s). How would you combine these quantities to yield an expression with dimensions of time? One with dimensions of velocity?

Core Concepts in College Physics Workbook

Solving the problem It helps to break the problem into manageable subproblems. For example, if we could write an expression for Bicyclist A's position x_A as a function of time, and another for Bicyclist B's position x_B as a function of time, then Bicyclist A must pass Bicyclist B at the instant when their positions are the same: $x_A = x_B$.

3. Both Bicyclist A and Bicyclist B move with constant acceleration. (In B's case, remember that zero is a constant!) What is the equation for the position as a function of time for an object with constant acceleration?

4. For both Bicyclist A and Bicyclist B, some of the quantities in the equation you've just written are zero. Dropping these terms, write simplified expressions for x_A and x_B.

5. To find the time t when they meet, set the expressions for x_A and x_B equal to each other. Solve the resulting equation for t. (Note: Do not replace algebraic symbols with numbers before solving for t. This is good advice for almost every problem; do not replace your equations with numbers until the last step of the problem.)

6. In your expression for t, is A's acceleration a_A in the numerator or the denominator? Where is B's velocity v_B? Can you give physical interpretations for each of these answers?

7. To find A's velocity when she passes B, write an expression for her velocity as a function of time. Then, into this expression, substitute for the time t when she catches him using the expression you found in step 5.

8. Except for dimensionless constants, do your answers to steps 5 and 7 resemble your predictions based on dimensional analysis from step 2?

PROBLEM 10 (SCREEN 3.4)

One-Dimensional Motion at Constant Acceleration

Problem Description

In 1865, Jules Verne proposed sending men to the moon by firing a space capsule from a 220-m long cannon with a final velocity of 10.97 km/s. What would have been the (assumed constant) acceleration experienced by the space travelers as they moved along the cannon barrel? How much time would it take them to reach the end of the cannon?

Before we begin...

1. Identify the given information:

   initial velocity (before launch) $v_0 =$

   final velocity (after accelerating along cannon barrel) $v_f =$

   distance traveled while accelerating $\Delta x =$

Solving the problem

2. We know the distance traveled by the astronauts under the constant acceleration and their initial and final velocities. We do not yet know how much time the acceleration lasted. Which of the three constant-acceleration kinematics equations is best suited to finding the acceleration under these conditions?

3. Solve the equation you have chosen for a.

Core Concepts in College Physics Workbook

4. Compare this acceleration to the free-fall gravitational acceleration, $g = 9.8$ m/s². Is Jules Verne's plan a plausible one? Would *you* ride this spaceship?

5. Which of the remaining constant-acceleration kinematics equations provides the most direct solution for t, the time the acceleration lasted?

6. Solve this equation for t.

7. If you want to check your answer for consistency, plug your answers for a and t into the kinematics equation that you have not yet used. It will tell you the distance traveled while accelerating. Does this match the length of the cannon as given in the problem?

PROBLEM 11   (SCREEN 3.5)

Projectile Motion

Problem Description

A place kicker kicks a football from a point 36.0 from the goal, hoping to clear the crossbar, which is 3.05 m above the ground. When kicked, the ball leaves the ground with a speed of 20.0 m/s at an angle of 53° above the horizontal. Does the ball clear the crossbar or fall below it? As it passes the crossbar, is the ball still rising or is it falling?

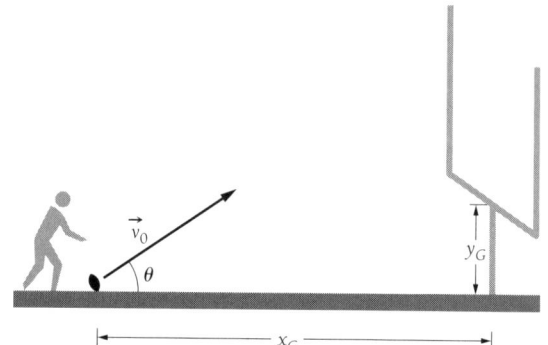

Before we begin...

Decide what coordinate axes you will use to describe the ball's motion. We will take the point where the ball is kicked as the origin, and call the direction from the origin toward the goal the +x direction. As usual, the +y direction will be straight upward.

1. Identify the given information:

   initial position of ball $x_0 =$ $y_0 =$
   initial speed of ball $v_0 =$
   initial angle of ball's trajectory to the horizontal $\theta =$
   position of goalpost (relative to the origin) $x_G =$ $y_G =$

2. Write the kinematics equations for freefall projectile motion near the earth's surface.

   $x =$ $v_x =$

   $y =$ $v_y =$

Solving the problem

The most important point to understand when solving projectile-motion problems is that the object's horizontal and vertical motion are independent of one another. The equations for horizontal motion (for x and v_x, as functions of time) are much simpler than those for vertical motion (for y and v_y). Thus, a good strategy is first to use the simpler horizontal motion to fill in missing information about velocities or time, and then tackle the more complicated vertical motion, armed with this new information. This problem will illustrate this technique.

3. Use trigonometry to break the initial velocity vector into its x and y components v_{x0} and v_{y0}.

4. Examine the equation for x as a function of time. At the moment the ball crosses the goalpost, x must be equal to x_G. At that moment, there should be only one unknown quantity in the equation. Solve for this unknown.

5. You have gained some information by considering the ball's horizontal motion. Turn now to the vertical motion, namely, the equation for y as a function of time. You should have enough information to find the ball's altitude y at the moment it crosses the goalpost. What is this value? Does the ball pass above or below the crossbar?

6. To determine whether the ball is rising or falling as it passes the goalpost, you must find whether v_y, the vertical component of its velocity, is positive or negative at that moment. Solve the equation for v_y. Is the ball rising or falling?

PROBLEM 12 (SCREEN 3.5)

Projectile Motion

Problem Description

A ball is thrown horizontally from the top of a building 35 meters high. The ball strikes the ground 80 m from the base of the building. Neglecting air resistance, find the time the ball is in flight. What was its initial velocity? Find the horizontal and vertical components of the velocity in the instant before the ball hits the ground.

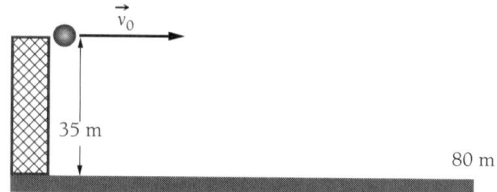

Before we begin...

Decide what coordinate axes you will use to describe the ball's motion. We will make the "standard" choice for projectile motion: the $+y$ direction is straight up, and the $+x$ direction is to the right, as shown in the diagram.

1. Identify the relevant quantities if given in the problem, or note them as unknown:

   distance traveled in the x direction $\Delta x =$
   distance traveled in the y direction $\Delta y =$
   x-component of initial velocity $v_{0x} =$
   y-component of initial velocity $v_{0y} =$
   freefall gravitational acceleration $g =$

2. Sketch the ball's flight on the diagram, labelling all known quantities.

Solving the problem

3. Since air resistance is neglected, the standard equations of projectile motion in a constant gravitational field apply. Write these four equations (for Δx, Δy, v_x, and v_y, as functions of time).

Core Concepts in College Physics Workbook

4. Which of these four equations is best suited for finding the ball's time of flight? (*Hint:* Find an equation in which the only unknown quantity is *t*.)

5. Solve this equation to find an expression for *t*. Substitute known values into this expression to find the flight time in seconds.

6. You already know the vertical component of the ball's initial velocity, v_{0y}. Which of the four projectile-motion equations is best suited for finding the horizontal component v_{0x}?

7. Solve this equation for v_{0x} and substitute known values.

8. What are the magnitude and direction of the ball's initial velocity vector $\vec{v}_0$?

9. You already know the time *t* at which the ball hits the ground. Use the remaining two projectile-motion equations to find v_x and v_y, the horizontal and vertical components of the ball's velocity at that instant.

PROBLEM 13 (SCREEN 3.6)

Uniform Circular Motion

Problem Description

A stone on the end of a string is whirled at constant speed around a horizontal circle of radius 0.30 m. The plane of the circle is 1.2 m above the ground. The string breaks, and the stone lands on the ground, 2.0 m away from the point directly beneath its position when the string broke. While the stone was still attached to the string and moving in a circle, what was its centripetal acceleration? Neglect air resistance.

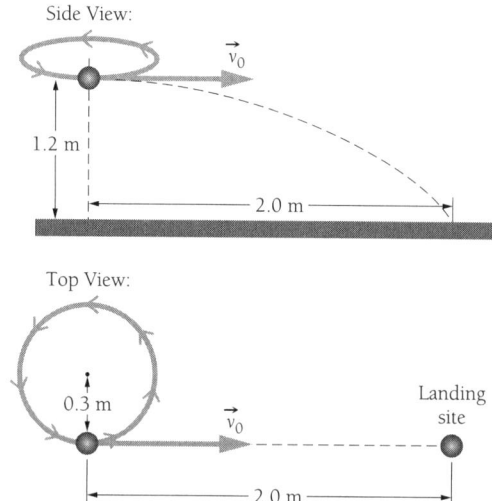

Before we begin... Decide what coordinate axes you will use to describe the stone's motion. We will make the "standard" choice for projectile motion: the $+y$ direction is straight up, and the $+x$ direction is to the right, as shown in the diagram.

1. Identify the relevant quantities if given in the problem, or note them as unknown:

   Before the string breaks:
   radius of the stone's circular path $r =$

   After the string breaks:
   distance traveled in the x direction $\Delta x =$
   distance traveled in the y direction $\Delta y =$
   freefall gravitational acceleration $g =$

2. This problem involves two well-known special types of two-dimensional motion. What are they?

Core Concepts in College Physics Workbook

3. What is the formula for the centripetal acceleration of an object moving in a circle? Which of the important quantities in this equation do you not yet know?

4. The key insight for this problem is that the stone's constant speed v during its circular motion must be equal to the initial speed v_{0x} of its projectile motion. Why must this be true?

Solving the problem

The insight described in step 4 suggests a strategy for breaking the problem into workable subproblems: we can use the given information about the stone's projectile motion to find its initial speed v_{0x}, then use this speed along with the known radius to find its centripetal acceleration during the circular motion.

5. The projectile motion phase of the stone's motion is essentially identical to the motion described in Problem 12. What are the important similarities between these two cases of projectile motion? Are there any important differences?

6. You now see one reason why you are encouraged to work out mathematical expressions for your answer, rather than immediately substituting given numbers. The equations you derived in steps 5 and 7 of Problem 12 apply here as well. Substitute the known quantities from this problem into those expressions to find the stone's speed v_{0x} at the instant of its release.

7. Use the equation you cited in step 2 to find the magnitude of the stone's centripetal acceleration vector. What is the direction of this vector?

PROBLEM 14   (SCREEN 3.7)

Relative Motion

Problem Description

A boat crosses a river of width $w = 160$ m in which the current has a uniform speed of 1.50 m/s. The pilot maintains a bearing (i.e., the direction in which the boat points) perpendicular to the river and a throttle setting to give a constant speed of 2.00 m/s relative to the water. What is the speed of the boat relative to a stationary shore observer? How far downstream from the initial position is the boat when it reaches the opposite shore?

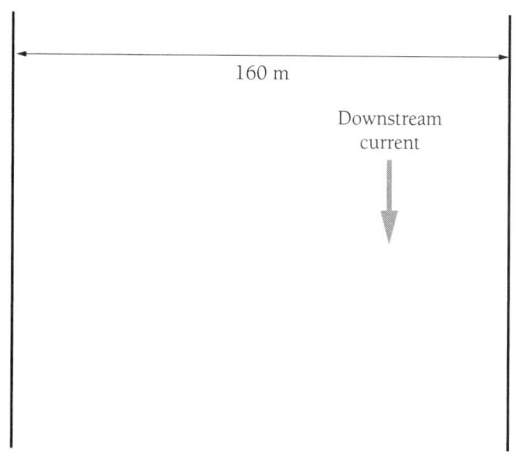

Before we begin...

1. In a diagram, show the velocity of the boat relative to the river, $\vec{v}_{br}$, and the velocity of the river relative to the shore, $\vec{v}_{rs}$. Indicate the vector sum $\vec{v}_{bs}$ of the two velocities. (Ignore the scale of the width of the river.)

2. Why is the vector sum important to this problem?

Solving the problem

Not only is the boat moving across the river due to the boat's motor, but it is moving downstream due to the river's current. A stationary observer on the shore will see the relative velocity due to both of these influences. In solving this problem, the concepts of vector addition and uniform motion will be employed.

3. Use the Pythagorean theorem to calculate the magnitude of the velocity of the boat relative to the shore $\vec{v}_{bs}$. This is the speed of the boat relative to the shore observer.

4. Since the heading of the boat is directly perpendicular to the current, the boat speed relative to the water will determine the time required for the crossing. Assuming that the river is perpendicular to the x axis, the boat will move across the river with $\vec{v}_{br}$ = 2.00 m/s. Calculate the time required to cross the river.

5. The entire time the boat is moving across the river, it is also moving downstream at the rate of 1.50 m/s. Use the appropriate equation of motion to calculate the distance downstream that the boat will travel.

MODULE 4 Forces

INTRODUCTION

In the previous module, we learned to describe motion elegantly and concisely. In this module, we begin to explore the forces that cause this motion. Newton's three laws of motion relate the net force acting on an object to its resulting motion. No force is necessary to keep a moving object in motion, but it does take a net force to make any object accelerate.

Because forces have both magnitude and direction, they are vector quantities.

DEFINITIONS

Newton's first law. An object at rest remains at rest and an object in motion continues in motion, at constant velocity, unless acted upon by a net external force.

Newton's second law. The net force acting on an object at any instant is equal to the object's mass multiplied by its instantaneous acceleration.

Newton's third law. When two bodies interact, the force exerted by body 1 on body 2 is equal in magnitude and opposite in direction to the force exerted by body 2 on body 1.

Weight. The force exerted on an object by the Earth's gravitational attraction.

Normal force. The force that occurs between objects in direct contact, which keeps them from falling through each other. The normal force is always directed perpendicular (normal) to the surface of contact.

Centripetal force. The force that accelerates a mass into circular motion.

USEFUL EQUATIONS

Newton's second law expressed in equation form is

$$\vec{F}_{net} = M\vec{a}$$

Weight is computed as mass times acceleration due to gravity:

$$\vec{w} = M\vec{g}$$

Newton's third law can be expressed in equation form by

$$\vec{F}_{12} = -\vec{F}_{21}$$

Centripetal force is calculated by

$$F_r = ma_r = m\frac{v^2}{r}$$

PROBLEM 15 (SCREEN 4.2)

Motion, Newton's First Law, and Force

Problem Description
Two people are pulling a boat through the water as shown in the figure. Each person exerts a force of 600 N directed at a 30.0° angle relative to the forward motion of the boat. If the boat moves with constant velocity, find the resistive force, $\vec{F}$, exerted on the boat by the water.

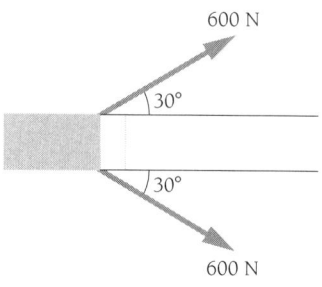

Before we begin...

1. If the boat moves at a constant velocity, what is the net force that is being exerted on the boat?

2. Label the forces that the people are exerting as $\vec{T}_1$ for the uppermost force and $\vec{T}_2$ for the lower oriented force. Draw the resistive force onto the diagram shown here.

Solving the problem

3. Because the boat is not accelerating, Newton's first law tells us that the sum of the forces acting upon the boat must equal zero. Selecting the boat to be a point mass at the origin of the cartesian coordinate system, resolve the forces into their x and y components. The x components must sum to zero, as must the y components.

4. Solve for the x and y components of the resistive force $\vec{F}$.

Core Concepts in College Physics Workbook

PROBLEM 16 (SCREEN 4.4)

Newton's Second Law

Problem Description

A constant force changes the speed of an 85-kg sprinter from 3.0 m/s to 4.0 m/s in 0.50 s. His direction does not change. Calculate the magnitude of the acceleration of the sprinter, the magnitude of the force, and the magnitude of the acceleration of a 58-kg sprinter experiencing the same force. (Assume linear motion.)

Before we begin...

1. Identify the relevant information:

   initial speed of sprinter $v_i =$
   final speed of sprinter $v_f =$
   mass of original sprinter $m_1 =$
   elapsed time $\Delta t =$
   mass of second sprinter $m_2 =$

Solving the problem

2. Using the given information from above and the equations for uniform acceleration from Module 3, compute the acceleration that the sprinter must experience.

3. If the acceleration of a given mass is known, which one of Newton's laws of motion allows you to calculate the net force acting upon the mass?

4. Use this information to find the force on the sprinter.

5. Finally, use this known force to calculate the unknown acceleration of the 58-kg sprinter.

PROBLEM 17 (SCREEN 4.5)

Newton's Third Law

Problem Description

A small bug is placed between two blocks of masses m_1 and m_2 ($m_1 > m_2$) on a frictionless table. A horizontal force, $\vec{F}$, can be applied to either m_1, as in figure (a), or m_2, as in figure (b). Show that the bug has a greater chance of surviving when the force is applied to m_1. (*Hint*: Determine the contact force in each case.)

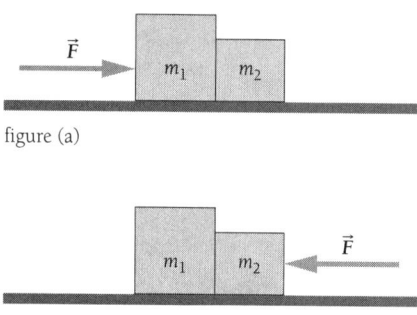

figure (a)

figure (b)

Before we begin...

1. Consider the situation in figure (a), but pretend we do not know which mass is larger. Consider the case where the force is on m_1, and draw the free-body diagrams for the two blocks.

2. What is the relationship of the accelerations of the two blocks?

Solving the problem

3. Using the free-body diagrams, apply Newton's second law to relate the net force on each block to the block's mass and acceleration. Take care to keep the appropriate signs to the forces.

4. Solve the two simultaneous equations for the acceleration, in terms of the applied force F, and the masses m_1 and m_2.

5. Substitute the expression for the acceleration into either of the second law equations, and solve to obtain an expression for the contact force. (*Hint:* Use the simpler of the two equations.)

6. Since your solution makes no assumption about the relative sizes of m_1 and m_2, you can immediately write down the contact force for figure (b). (*Hint:* What would happen if you swapped m_1 and m_2 in figure (a)?)

7. With $m_1 > m_2$, compare your expressions for the contact force in each case.

PROBLEM 18 (SCREEN 4.5)

Newton's Third Law

Problem Description
Three blocks are in contact with each other on a frictionless, horizontal surface. A horizontal force $\vec{F}$ is applied to M_1. If $M_1 = 2.00$ kg, $M_2 = 3.00$ kg, $M_3 = 4.00$ kg, and $F = 18.0$ N, draw free-body diagrams of each block and find the acceleration of the blocks. What is the *resultant* force on each block? What are the magnitudes of the contact forces between the blocks?

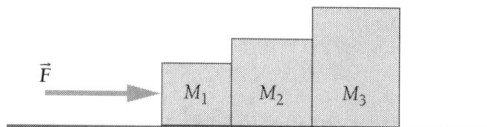

Before we begin... 1. Draw the free-body diagrams for the three blocks.

2. What is the relationship of the acceleration of each block to the accelerations of the other blocks?

Solving the problem 3. Using the free-body diagrams, apply Newton's second law to relate the net force on each block to the block's mass and acceleration. Take care to keep the appropriate signs to the forces.

4. Solve the three simultaneous equations for the acceleration.

5. Substitute the value of the acceleration into each of the second law equations to evaluate the contact force acting upon the object. (*Hint:* Begin with either Block 1 or Block 3, not Block 2.)

Core Concepts in College Physics Workbook

PROBLEM 19 (SCREEN 4.6)

Free-Body Diagrams

Problem Description
Two blocks on a frictionless horizontal surface are connected by a light string as in the figure, where $m_1 = 10$ kg and $m_2 = 20$ kg. A force of 50 N is applied to the 20-kg block. Determine the acceleration of each block and the tension in the string. Repeat the problem for the case in which the coefficient of kinetic friction between each block and the surface is 0.10.

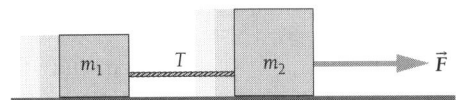

Before we begin...

1. What is the relationship between the accelerations of the blocks?

2. Draw free-body diagrams for each of the blocks. Consider the positive direction for each block to be the direction of motion.

Solving the problem

3. From the information above and the free-body diagrams, write Newton's second law as it applies to each of the blocks.

4. Solve the equations simultaneously to find the acceleration of the system. This must be the same value for each of the blocks.

(continued on next page . . .)

Module 4 **Forces**

5. Substitute back into the two equations to evaluate the tension in the string.

6. To include the coefficient of kinetic friction, add the kinetic friction force to the free-body diagrams from above.

7. Does the relationship of the accelerations of the blocks change if you include the coefficient of kinetic friction?

8. Modify your equation for Newton's second law to include the coefficient of kinetic friction.

9. Again, solve the equations simultaneously to find the acceleration of the system.

10. Substitute back into the two equations to evaluate the tension in the string, this time including the coefficient of kinetic friction.

PROBLEM 20   (SCREEN 4.6)

Free-Body Diagrams

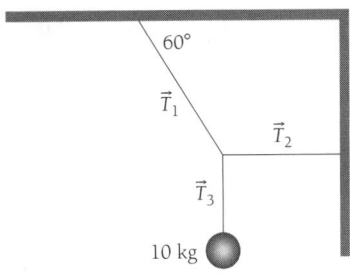

Problem Description
Find the tension in each cord of the system illustrated here.

Before we begin...

1. Draw two free-body diagrams for the problem. Include all forces acting upon the junction of the string in the first diagram and all forces acting upon the ball in the second.

2. Is the system in equilibrium, or does a net force act upon the object?

3. What is the weight of the ball?

Solving the problem

4. Because the system is in equilibrium, the net force acting upon the ball must be zero. From the second free-body diagram, compute the magnitude of the tension, $\vec{T}_3$.

5. The first free-body diagram shows that the tension $\vec{T}_3$ is pulling vertically downward ($-y$ direction). What are the directions of $\vec{T}_1$ and $\vec{T}_2$?

    Since the knot is in equilibrium, $\vec{T}_1 + \vec{T}_2 + \vec{T}_3 = 0$. This requires that simultaneously:

    $$T_{1x} + T_{2x} + T_{3x} = 0 \quad \text{and} \quad T_{1y} + T_{2y} + T_{3y} = 0$$

6. Resolve the tensions into their x and y components and solve the two equations simultaneously.

Module 4 Forces

PROBLEM 21 (SCREEN 4.6)

Free-Body Diagrams

Problem Description
Three objects are connected by strings as shown in the figure. The masses are 4.0 kg, 1.0 kg, and 2.0 kg, respectively, and both the table and the pulleys are frictionless. Determine the acceleration of each object and their directions. Determine the tensions in the two cords.

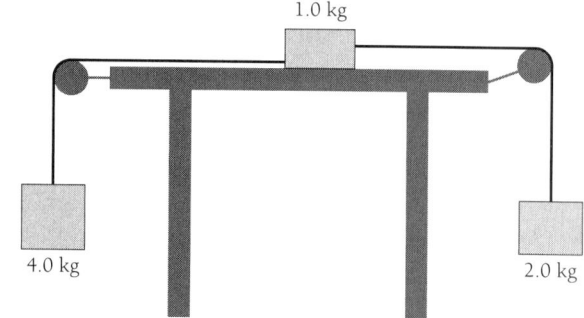

Before we begin...

1. Which object do you predict will accelerate in the vertically downward direction, the 4.0-kg object or the 2.0-kg object?

2. What is the relationship between the accelerations of the objects?

3. Draw free-body diagrams for each of the objects. Consider the positive direction for each object to be the direction you have predicted it will move.

Solving the problem

4. From the information above and the free-body diagrams, write Newton's second law as it applies to each of the three objects.

5. Solve the equations simultaneously to find the acceleration of the system. This must be the same value for each of the objects.

6. Substitute back into the several equations to evaluate the tensions in the two cords.

PROBLEM 22 (SCREEN 4.6)

Free-Body Diagrams

Problem Description
On takeoff, the combined action of the engines and wings of an airplane exerts an 8000-N force on the plane, directed upward at an angle of 65.0 degrees above the horizontal. The plane rises with constant velocity in the vertical direction while continuing to accelerate in the horizontal direction. What is the weight of the plane? What is its horizontal acceleration?

Before we begin...

1. Draw a free-body diagram for the plane.

2. What directions seem most convenient for the x and y axes?

Solving the problem

3. Find the components of the forces along each of the axes chosen above.

(continued on next page . . .)

4. Write Newton's second law in component form. Since the plane is not accelerating in the vertical direction, the vertical forces must sum to zero; the weight must be equal in magnitude to the vertical component of the 8000-N force.

5. Divide the weight by g, the acceleration of gravity, to find the mass of the plane. Solve the horizontal equation for acceleration, and substitute the value for the mass to find the value for its horizontal acceleration.

PROBLEM 23 (SCREEN 4.7)

Centripetal Force

Problem Description
An automobile moves at a constant speed over the crest of a hill. The driver moves in a vertical circle of radius 18.0 m. At the top of the hill, she notices that she barely remains in contact with the seat. Find the speed of the vehicle.

Before we begin...

1. Draw a free-body diagram for the driver as she crests the hill. Does a normal force exist between the driver and the seat?

2. What force is providing the centripetal acceleration that she is experiencing?

Solving the problem

3. Since the driver is moving at a constant speed in a vertical circle, she must have a centripetal force acting toward the center of the circle. Write the expression for the net force providing the centripetal force and set it equal to centripetal force.

4. Solve this for the speed of the car.

5. Why does the mass of the driver not matter in this problem?

Module 4 **Forces**

PROBLEM 24 (SCREEN 4.8)

Fictitious Forces: Motion in Accelerated Reference Frames

Problem Description

A 5.00-kg mass attached to a spring scale rests on a frictionless, horizontal surface as in the figure. The spring scale, attached to the front end of a boxcar, reads 18.0 N when the car is in motion. If the scale reads zero when the car is at rest, determine the acceleration of the car while it is in motion. What will the spring scale read if the car moves with constant velocity? Describe the forces on the mass as observed by someone in the car and by someone at rest outside the car.

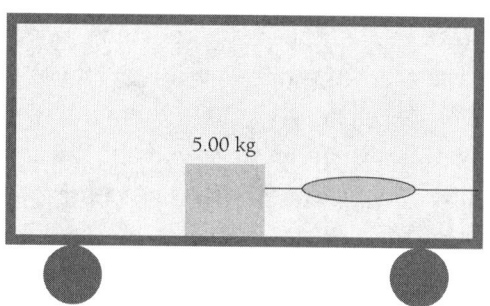

Before we begin...

1. Draw a free-body diagram for the mass as observed from the ground. Assume that the spring scale exerts a tension $\vec{T}$ directed in the $+x$ direction upon the mass; this tension is measured by the scale.

Solving the problem

2. The boxcar is accelerating in the $+x$ direction. Since there is no friction, the net force exerted on the mass m is the tension $\vec{T}$. Calculate the acceleration of the mass with respect to the inertial frame of reference (a stationary point on the ground).

3. If the car moves with constant velocity, what is its acceleration? Use this information to calculate the tension under this condition.

4. In the noninertial frame of reference (within the boxcar), the mass appears to be in equilibrium. What is the net force acting upon an object in equilibrium? Use this information to compute the fictitious force in the accelerated frame of reference.

MODULE 5

Work and Energy

INTRODUCTION

The concepts of work and energy and the law of conservation of energy are important in examining mechanical systems. These concepts build upon those of Module 4, *Forces* (and Newton's laws of motion). Using the work-energy relationships often simplifies the analysis of a situation, in contrast to a direct application of Newton's second law. Work, energy, and energy conservation are important concepts in all areas of physics.

In this module, we use the scalar product of two vectors to derive a scalar quantity.

DEFINITIONS

Work. The product of the component of force in the direction of displacement and the magnitude of the displacement.

Kinetic energy. The energy that an object has by virtue of its motion.

Conservative force. If the work done by a force on a moving object is independent of the path that the object takes from its start to its finish position, then the force is conservative. Gravity is one example of a conservative force.

Potential energy. The energy that an object has by virtue of its state, shape, or position. It is equal to the work that would be done by all conservative forces as the object moved into some arbitrary "zero-energy" state.

Power. A measure of the rate at which work is done or energy is used.

Conservation of energy. A law stating that energy cannot be created or destroyed. Energy can only be converted among the three forms of energy: kinetic, potential, and thermal energy.

USEFUL EQUATIONS

The work done by a constant force is computed by the equation

$$W = \vec{F} \cdot \vec{s}$$

where $\vec{F}$ is the applied force and $\vec{s}$ is the displacement.

If the force varies over the displacement, use the equation

$$W_{net} \sum_i \vec{F}_i \cdot \Delta \vec{s}_i = \sum F \Delta s \cos \theta$$

where the displacements $\Delta \vec{s}$ are each small enough that the force is very nearly constant over them.

The work-energy theorem can be expressed by the equation

$$W_{net} = \Delta K$$

where K is the kinetic energy

$$K = \frac{1}{2} mv^2$$

Potential energy U of an object at a location is defined as the negative of the work done by a conservative force as the object is moved from an arbitrary "zero point" to that location.

$$\Delta U = -W_c = -\sum \vec{F}_c \cdot \Delta \vec{s}$$

In a constant gravitational acceleration g, the potential energy is

$$U_{grav} = mgh$$

where h is the object's height above the "zero point."

Power is computed by the equation

$$P = \frac{\Delta W}{\Delta t} = \vec{F} \cdot \vec{v}$$

Module 5   **Work and Energy**

PROBLEM 25 (SCREEN 5.2)

Work

Problem Description

Starting from rest, a 5.0-kg block slides 2.5 m along a rough 30° incline in 2.0 s. Three forces act on the block during its descent: its weight $m\vec{g}$, the normal force $\vec{N}$, and friction $\vec{f}$ between the ramp and the block (which you may assume is constant). Calculate the work done on the block by each of these three forces.

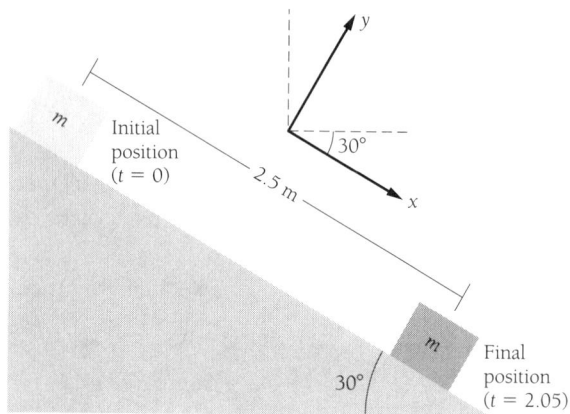

Before we begin...

Choose the coordinate system you will use to describe the block's motion. We will take the x axis to point downhill along the ramp's surface, and the y axis to be perpendicular to the surface, as shown in the diagram.

1. Identify the relevant information:

   mass of the block $m =$
   angle of incline $\theta =$
   distance moved by the block $\Delta s = \Delta x =$
   initial speed of block $v_0 =$
   time over which motion occurred $t =$

2. Sketch the directions of the three force vectors and the displacement $\Delta \vec{s}$ on the above diagram.

Solving the problem

3. Since work is the dot product of a force and a displacement, it depends on the angle between the force and the displacement. What are the angles between the block's displacement and the three force vectors?

   (angle between $\Delta \vec{s}$ and $m\vec{g}$) =
   (angle between $\Delta \vec{s}$ and $\vec{N}$) =
   (angle between $\Delta \vec{s}$ and $\vec{f}$) =

Core Concepts in College Physics Workbook

4. You now have enough information to calculate the work done on the block by its weight $m\vec{g}$ and by the normal force $\vec{N}$. What are these results? (Why don't you need to know the magnitude of $\vec{N}$?)

We now want to know the magnitude of the friction $\vec{f}$. From the information given in the problem, it's possible to find the block's acceleration directly, and then work backward from Newton's second law to deduce the frictional force. Since all forces are assumed constant, we can use the kinematics equations for constant acceleration.

5. Which kinematics equation is best suited to finding the acceleration $\vec{a}$ from the given information? Apply this equation and solve for $\vec{a}$.

6. Write expressions for the x and y components for all the forces and the acceleration. (*Hint*: You may want to sketch each vector in relation to the coordinate axes.)

7. Apply Newton's second law and solve for the unknown force $\vec{f}$. (*Hint*: You won't need to apply Newton's second law to both the x and y components of the motion. Which component should you concentrate on? This is one of the benefits of the "tilted" coordinate axes we've chosen for this problem.)

8. What is the work performed on the block by the friction $\vec{f}$?

9. It is possible to bypass steps 5 through 8 and find the work done by $\vec{f}$ much more easily using the relationship between work and energy. Can you figure out how? (If you haven't yet studied the work-energy theorem, revisit this problem once you're familiar with it!)

Module 5 **Work and Energy**

PROBLEM 26 (SCREEN 5.3)

Important Examples of Work:
Gravity and Springs

Problem Description
A cheerleader lifts his 50.0-kg partner straight up off the ground a distance of 0.60 m before releasing her. If he does this 20 times, how much work has he done?

Before we begin...

1. How much force does the cheerleader have to exert to lift his partner each time?

2. Through what displacement is this average force exerted each time?

Solving the problem

3. Employ the equation for work done by a constant force to calculate the work done for each lift.

4. Because the process was done 20 times, multiply the work done for a single lift by 20.

Core Concepts in College Physics Workbook

PROBLEM 27 (SCREEN 5.5)

Energy

Problem Description
The driver of a 1000-kg car is pulled over for speeding. "That vehicle's got as much kinetic energy as mine did. Why didn't you pull him over too?" he complains, pointing out a 4000-kg truck moving at 60 miles per hour. If the driver is correct that the two vehicles had the same kinetic energy, how fast was he driving?

Before we begin...

1. Identify the relevant information:

   mass of the car m_{car} =
   mass of the truck m_{truck} =
   speed of the truck v_{truck} =

   (*Note*: While it is generally a good idea to translate nonstandard units such as miles per hour into meters per second, it is not necessary in this problem. Why?)

Solving the problem

2. What is the formula for an object's kinetic energy?

3. Translate the driver's claim that the vehicles had the same kinetic energy into an equation relating their masses and speeds.

4. Solve this equation for the driver's speed v_{car}.

5. Instead of writing and solving equations, can you make a simple argument in words, based on your understanding of kinetic energy, leading to the same answer?

Module 5 **Work and Energy**

PROBLEM 28 (SCREEN 5.6)

Conservative Forces

Problem Description

The floor of a third-story apartment is 6.0 m above the street. A cat of mass 4.0 kg jumps from the floor onto a 2.0-m high bookshelf. What was the cat's gravitational potential energy, before and after she jumped onto the shelf, measured relative to the floor? Measured relative to the street? How can potential energy be a useful concept if its value changes depending on which reference point you choose?

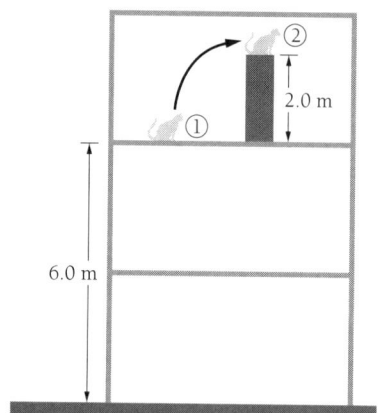

Before we begin...

1. Identify the relevant information:

   mass of the cat $m =$
   gravitational acceleration $g =$
   height of bookshelf above floor $h_{shelf} =$
   height of floor above street $h_{floor} =$

Solving the problem

2. Relative to the floor, what are the cat's altitude h_1 before she jumps and her altitude h_2 after she jumps onto the shelf?

3. What is the formula for gravitational potential energy in a constant or nearly-constant gravitational field?

4. Applying the formula from step 3, find the cat's potential energy relative to the floor U_1 (before jumping) and U_2 (after jumping).

5. Relative to the street, what are the cat's altitude h_1 before she jumps and her altitude h_2 after she jumps onto the shelf?

6. Applying the formula from step 3, find the cat's potential energy relative to the street U_1 (before jumping) and U_2 (after jumping).

7. How do you interpret the difference in your answers to steps 4 and 6? What aspect of your answer did not change when you shifted your reference point from the floor to the street?

PROBLEM 29   (SCREEN 5.7)

The Work-Energy Theorem

Problem Description

The launching mechanism of a toy gun consists of a spring of unknown spring constant, as shown at right. If its spring is compressed a distance of 12.0 cm, the gun can launch a 20.0-g projectile to a maximum height of 20.0 m when fired vertically from rest. Neglecting friction and air resistance, what is the spring constant k? How fast is the projectile moving as it passes through the spring's equilibrium position?

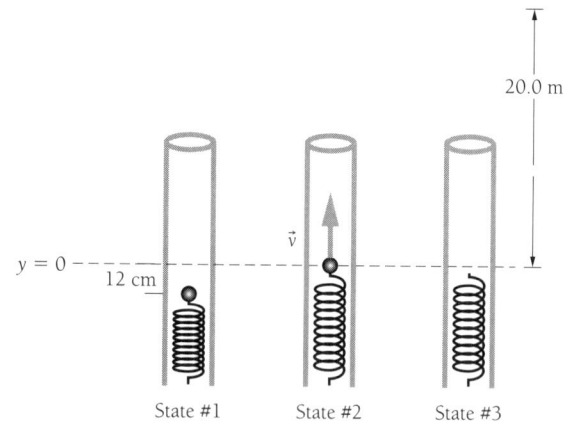

Before we begin...

This problem involves gravitational potential energy, which must be calculated relative to some reference point (where it is assumed to be zero). We will choose as this reference point the spring's equilibrium position, which we label $y = 0$.

When solving problems using energy conservation, be sure to clearly identify the two or more important or revealing states that the system passes through. In this problem, the three most interesting points in the projectile's history are:

> State #1: The spring is compressed but the gun has not yet been fired.
> State #2: The projectile is passing through the equilibrium position ($y = 0$) on its way upward.
> State #3: The projectile has reached the highest point in its flight.

1. Identify the relevant information:

   mass of the projectile $m =$
   height of the projectile before firing $y_1 =$
   height of the projectile passing equilibrium $y_2 =$
   height of the projectile at top of flight $y_3 =$

   (Remember to convert all quantities to standard units of kg, m, and s.)

Core Concepts in College Physics Workbook

Solving the problem

We will first compare the system's energy before firing (State #1) and at the top of its flight (State #3). (State #1 is interesting because it involves a crucial piece of missing information—the spring constant k of the compressed spring. State #3 is interesting because it is the only state where we have complete knowledge of the system: a projectile of known mass is momentarily at rest a known distance above the ground.)

2. When the spring is compressed and ready to fire (State #1), what forms of energy are present? Write an expression for the system's total energy E_1 in this state.

3. When the projectile has reached its maximum height (State #3), is it moving? What forms of energy are present at that instant? Write an expression for the system's total energy E_3 in this state.

4. Was any work done by nonconservative forces as the system evolved from State #1 to State #3? Based on your answer to this question, should E_1 be greater than, less than, or equal to E_3?

5. Combine the energy expressions from steps 2 and 3 into an equation expressing the claim you made in step 4.

6. Solve this equation for the spring constant k.

(continued on next page . . .)

The projectile's speed v as it passes equilibrium can be found by comparing its energy in State #2 to that of either of the other two states. For simplicity, we choose to compare it to State #3, where only one form of energy is present.

7. As the projectile passes the equilibrium point (State #2), what forms of energy are present? Write an expression for the system's total energy E_2 in this state.

8. Should E_2 be greater than, less than, or equal to E_3? Combine the expressions for E_2 and E_3 into a single equation, and solve this equation for the particle's speed v, the magnitude of the vector $\vec{v}$, as it passes equilibrium.

PROBLEM 30 (SCREEN 5.7)

Work-Energy Theorem

Problem Description
An 80.0-kg firefighter slides down a 6.0-meter high pole. As he descends, the pole exerts on him a constant upward frictional force $\vec{f}$. He starts from rest at the top of the pole, and when he reaches the bottom, he is moving downward at 5.0 m/s. How much mechanical energy was "lost" (i.e., transformed to heat)? What is the magnitude of the frictional force $\vec{f}$?

Before we begin...

This problem involves gravitational potential energy, which must be calculated relative to some reference point (where it is assumed to be zero). We will choose as this reference point the bottom of the pole, which we label $y = 0$.

When solving problems using energy conservation, be sure to clearly identify the two or more important or revealing states that the system passes through. In this problem, the two most interesting points in the firefighter's history are:

> State #1: The firefighter begins to slide from rest at the top of the pole.
>
> State #2: The firefighter, still moving downward, is just reaching the bottom of the pole.

1. Identify the relevant information:

> mass of the firefighter $m =$
> position of the firefighter before sliding down $y_1 =$
> position of the firefighter at end of descent $y_2 =$
> speed of the firefighter before sliding down $v_1 =$
> speed of the firefighter at end of descent $v_2 =$

(continued on next page . . .)

Module 5 **Work and Energy**

Solving the problem

2. When the firefighter begins his descent (State #1), what forms of energy are present? Write an expression for the system's total energy E_1 in this state.

3. When the firefighter is about to reach the bottom of the pole (State #2), what forms of energy are present? Write an expression for the system's total energy E_2 in this state.

4. If you have not already done so, evaluate E_1 and E_2 in Joules. How much energy was "lost" during the descent?

5. Does your answer to step 4 violate the law of conservation of energy? Why did the system's mechanical energy change? What happened to the "missing" energy?

6. Based on your answers to steps 4 and 5, how much work W_{n-c} was done on the firefighter by the nonconservative force $\vec{f}$?

7. Using the definition of the work done by a constant force on an object moving through a known displacement, write an expression for W_{n-c} in terms of f and Δy.

8. Combine your answers to steps 6 and 7 into a single equation, and solve for f.

PROBLEM 31 (SCREEN 5.8)

Power

Problem Description
A 650-kg elevator starts from rest. It moves upward for 3.00 s with constant acceleration until it reaches its cruising speed, 1.75 m/s. What is the average power of the elevator motor during this period? How does this power compare with its power while it moves at its cruising speed?

Before we begin...

1. Draw a free-body diagram of the forces acting upon the elevator.

2. Identify the given information:

    mass of elevator $m =$
    initial velocity $v_0 =$
    final velocity $v_f =$
    elapsed time $\Delta t =$

3. During the 3.00 seconds, the work done by the elevator motor is going to change what two types of energy for the elevator?

Solving the problem

4. Since both kinetic energy and potential energy are changed during the acceleration interval, the work done by the motor during this time equals the sum of these changes. Given the initial and final speeds, compute the change in kinetic energy.

(continued on next page . . .)

5. To compute the change in potential energy, we need to know the height Δy to which the elevator rises in 3.00 s. Use the kinematic equations to calculate this height. After finding Δy, compute the change in potential energy.

6. Calculate the average power as the work done by the motor divided by the time required.

Instantaneous power can be calculated by the relation

$$P = \vec{F} \cdot \vec{v}$$

7. What force is required to keep the elevator moving at a constant velocity? Use this force in the power equation to evaluate the cruising speed power requirement.

PROBLEM 32   (SCREEN 5.8)

Power

Problem Description
A car of mass 1.50×10^3 kg accelerates uniformly from rest to 15.0 m/s in 3.00 s. Neglecting all forms of friction and air resistance, find the average power delivered by the car's engine during the 3-s interval. Also, find the instantaneous power being delivered 2.00 seconds after the car begins to move.

Before we begin...   As usual in energy problems, it is important to distinguish carefully between the car's initial state, at $t = 0$, and its final state, at $t = 3.0$ s.

1. Identify the relevant information:

   mass of the car $m =$
   initial speed of the car $v_0 =$
   final speed of the car $v_f =$
   elapsed time over which the speed changed $\Delta t =$

2. What is the definition of power? Does this suggest a subproblem which should be solved before you can calculate the power?

Solving the problem   3. What were the car's initial and final kinetic energies?

4. How much work did the engine do on the car during the three seconds of acceleration? (*Hint:* Use your answer to step 3 and the work-energy theorem.)

(continued on next page . . .)

Module 5   **Work and Energy**

5. What was the average power delivered by the engine during the time of acceleration?

6. In real life, where friction and air resistance are important, would the engine need to deliver more or less power than you have just calculated, in order to accelerate the car to the same speed in the same amount of time? Why? Does the work-energy theorem not apply in the presence of friction?

We have now found the average power delivered to the car over the 3-s interval, but this is not necessarily the same as the power being delivered at any particular instant. Our next task, then, is to find the power being delivered at $t = 2.0$ s after the car begins to accelerate.

7. What is the formula for the power being delivered to an object at a given instant when it is subject to a force $\vec{F}$ and moving at velocity $\vec{v}$?

8. The problem states that the car's acceleration a is constant over the 3-s interval. Find the value of a.

9. Using the acceleration you have just determined, find the force F acting on the car and the car's speed v at $t = 2.0$ s.

10. At $t = 2.0$ s, how much power is being delivered to the car?

PROBLEM 33 (SCREEN 5.9)

Conservation of Energy

Problem Description
At time t_i, the kinetic energy of a particle is 30 J and its potential energy is 10 J. At some later time t_f, its kinetic energy is 18 J. If only conservative forces act on the particle, what are its potential energy and its total energy at time t_f? If the potential energy at time t_f is 5 J, are there any non-conservative forces acting on the particle? Explain.

Before we begin...

1. State the law of conservation of energy.

2. Identify the given information:
    initial kinetic energy $K_i =$
    initial potential energy $U_i =$
    final kinetic energy $K_f =$

Solving the problem

3. Evaluate the total energy E at time t_i.

(continued on next page . . .)

Module 5 **Work and Energy**

4. Applying the law of conservation of energy for a conservative system, compute the total energy E_f and the potential energy U_f at t_f.

5. If the final potential energy is also known, then the total mechanical energy can be computed independently for times t_i and t_f. What does the law of conservation of energy say if the mechanical energy changes between times t_i and t_f?

MODULE 6

Linear Momentum

INTRODUCTION

In this module, we introduce the law of conservation of linear momentum. We study the general form of Newton's second law of motion as well as collisions between objects. Finally, we learn to express the motion of a system of objects in terms of the center of mass of the system.

DEFINITIONS

Linear momentum, $\vec{p}$. The mass of an object multiplied by its velocity. Linear momentum is a vector quantity.

General form of Newton's second law. The net force acting upon a system is equal to the rate at which the linear momentum of the system changes with time.

Impulse, I. The change in momentum of the system. The product of the force and the time for which it acts.

Elastic collision. A process that conserves linear momentum and mechanical energy.

Inelastic collision. A process that conserves linear momentum but does not conserve mechanical energy.

Center of mass. A point that, in terms of mechanical behavior, would move as the system moves, if the point were to contain the entire mass of the system. The linear momentum of the center of mass is the sum of the linear momenta of all objects in the system.

USEFUL EQUATIONS

The equation for computing the linear momentum of a particle is

$$\vec{p} = m\vec{v}$$

If the net force applied to a system is zero, then linear momentum is conserved and constant.

$$\Delta \vec{p} = 0$$

The general form of Newton's second law of motion is

$$\vec{F}_{net} = \frac{\Delta \vec{p}}{\Delta t}$$

The impulse of a system is computed by the equation

$$\vec{I} = \vec{F} \Delta t = \Delta \vec{p}$$

If $\vec{F}$ varies with time,

$$\vec{I} = \langle \vec{F} \rangle \Delta t = \Delta \vec{p}$$

For a system of particles, the center of mass r_{CM} is calculated by the equation

$$\vec{r}_{CM} = \frac{\sum m_i \vec{r}_i}{M}$$

The momentum of the system can be expressed in terms of the center of mass by the relationship

$$\vec{p}_{total} = M\vec{v}_{CM}$$

The acceleration of the center of mass is given by Newton's second law:

$$\vec{a}_{CM} = \frac{\sum \vec{F}_{net}}{M}$$

PROBLEM 34 (SCREEN 6.2)

Conservation of Momentum

Problem Description
A 60.0-kg astronaut is on a space walk away from the shuttle when her tether line breaks. She is able to throw her 10.0-kg oxygen tank away from the shuttle with a velocity of 12.0 m/s to propel herself back to the shuttle. Assuming that she starts from rest (relative to the shuttle), determine the maximum distance she can be from the craft when the line breaks and still return within 60.0 s (the amount of time she can hold her breath).

Before we begin...

1. Choose an appropriate reference frame.

2. Identify the given information:

   $m_{astronaut} =$ $m_{tank} =$ $\Delta t =$

   $v_{astronaut\text{-}initial} =$ $v_{tank\text{-}initial} =$ $v_{tank\text{-}final} =$

3. What quantity must be first solved to determine the distance the astronaut can travel in 60 seconds?

4. State the general form of Newton's second law as expressed in terms of momentum.

(continued on next page . . .)

5. What is the net force applied to the system consisting of the astronaut and her oxygen tank?

6. What is conserved for the system during the interaction? How is that expressed mathematically?

Solving the problem

7. Write the expressions for the momentum of the system before and after the astronaut throws the tank.

8. Substitute these expressions in the expression of the conservation law from above, and solve for the unknown quantity.

9. Use the expression for the previously unknown quantity to solve for the distance the astronaut can be from the spacecraft.

PROBLEM 35   (SCREEN 6.4)

Impulse

Problem Description

The force F_x in Newtons acting on a 2.0-kg particle varies in time as shown. Find the impulse of the force, the final velocity of the particle if it is initially at rest, and its final velocity if it is initially moving along the x axis with a velocity of −2.0 m/s. What is the average force exerted on the particle for the time interval $t_i = 0$ to $t_f = 5.0$ s?

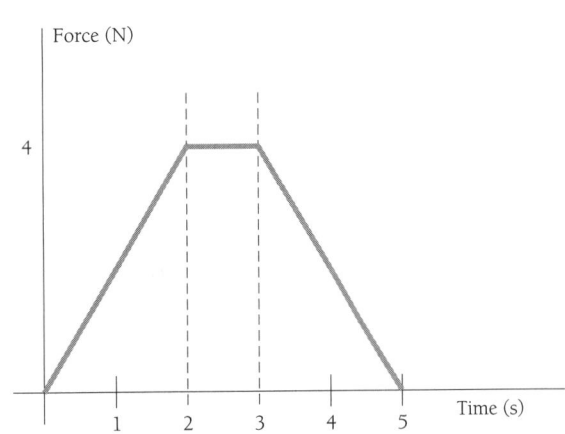

Before we begin...

1. Identify the given information:

    $m =$ v_i(particle initially at rest) $=$ v_i(particle initially moving) $=$

2. How is impulse related to force and the time interval over which the force is applied?

3. Recall that impulse equals the change in momentum. How is impulse related to mass and velocity?

Solving the problem

4. Since the force is not constant over the entire interval, you will need to employ the relation

    $$\Delta \vec{p} = \sum (\langle F_i \rangle \Delta t_i)$$

    for each of the three regions (two triangles and a rectangle) indicated in the graph.

(continued on next page . . .)

5. Write the equation for final momentum in terms of the initial momentum and the impulse.

6. Evaluate the equation for the given information about mass and initial velocity.

7. Show how large a constant force instead of a variable force would be required to act over the same 5 s time period in order to give the same value of impulse according to the relation

$$\Delta \vec{p} = \langle \vec{F} \rangle \Delta t$$

PROBLEM 36 (SCREEN 6.5)
Perfectly Inelastic Collisions

Problem Description
A 90-kg halfback running north with a speed of 10 m/s is tackled by a 120-kg opponent running south with a speed of 4.0 m/s. If the collision is perfectly inelastic and head-on, calculate the speed and direction of the players just after the tackle and determine the energy lost as a result of the collision. Account for the missing energy.

Before we begin...

1. Draw a sketch, using blocks to represent the two players. Label the masses and the velocities before the collision. Assign the positive direction.

2. Identify the given information:

   $m_1 =$ \qquad $m_2 =$ \qquad $\vec{v}_1 =$ \qquad $\vec{v}_2 =$

3. What is meant by a perfectly inelastic collision?

4. Is momentum conserved in this collision?

(continued on next page . . .)

Solving the problem

5. To solve the problem in which momentum is conserved but mechanical energy is not, evaluate the total momentum of the system before the collision.

6. Because the momentum after the collision is the same as before the collision, the velocity of the two players (who stick together and move as one mass) can now be evaluated.

7. Evaluate the total kinetic energy before the collision and compare it with the total afterward. Are they the same? Why or why not?

PROBLEM 37 (SCREEN 6.5)

Perfectly Inelastic Collisions

Problem Description
During the battle of Gettysburg, the gunfire was so intense that several bullets collided in midair and fused together. Assume a 5.0-g Union musket ball moving to the right at 250 m/s, 20° above the horizontal, and a 3.0-g Confederate ball moving to the left at 280 m/s, 15° above the horizontal. Immediately after they fuse together, what is their velocity?

Before we begin...

1. Draw a sketch of the system just before the collision. Label the objects and assign the positive and negative directions.

2. Identify the given information:

   $m_1 =$ $m_2 =$ $\vec{v}_1 =$ $\vec{v}_2 =$

3. What kind of collision do the two bullets undergo?

4. Is this problem solved in one or in two dimensions?

(continued on next page . . .)

Solving the problem

5. The perfectly inelastic collision takes place in two dimensions. This requires that the momentum in the x direction be conserved, as is the momentum in the y direction. Find the x and y components of momentum for each bullet before the collision.

6. Set the total x momentum before the collision equal to the combined mass multiplied by the x component of velocity after the collision. Solve for $\vec{v}_x$.

7. Repeat the above procedure for the y component.

8. Knowing the x and y components of velocity, find its magnitude and direction.

PROBLEM 38 (SCREEN 6.6)

Perfectly Elastic Collisions

Problem Description

Two blocks of mass $m_1 = 2.00$ kg and $m_2 = 4.00$ kg are released from a height of 5.00 m on a frictionless track as shown. The blocks undergo an elastic head-on collision. Determine the two velocities just before the collision. Determine the two velocities immediately after the collision. Determine the maximum height to which each block rises after the collision.

Before we begin...

1. Identify the given information:

   $m_1 =$ $m_2 =$ $\Delta h =$

   $v_1 =$ $v_2 =$

2. Considering that the track is frictionless, what happens to the gravitational potential energy of the blocks as they slide to the bottom of the incline?

3. In an elastic collision, what two quantities are conserved?

Solving the problem

4. Use the law of conservation of energy to solve for the speeds of the two blocks immediately before the collision. Assign the appropriate sign to each of these to convert them to velocities. (Let movement toward the right be positive.)

(continued on next page . . .)

5. Evaluate the total linear momentum and kinetic energy before the collision. Equate these values to the expressions for total linear momentum and kinetic energy after the collision, and solve for the velocities after the collision.

6. You can now employ the law of conservation of energy for each block after the collision to compute the height to which each will rise.

PROBLEM 39   (SCREEN 6.6)
Perfectly Elastic Collisions

Problem Description
Two shuffleboard disks of equal mass, one orange and the other yellow, are involved in a perfectly elastic glancing collision. The yellow disk is initially at rest and is struck by the orange disk moving initially to the right at 5.00 m/s. After the collision, the orange disk moves in a direction that makes an angle of 37.0° with its initial direction, and the velocity of the yellow disk is perpendicular to that of the orange disk (after the collision). Determine the speed of each disk after the collision.

Before we begin...

1. What quantities are conserved in an elastic collision?

2. Write down the equations which express these conservation laws.

3. Draw the relationship of the initial and final momentum vectors.

(continued on next page . . .)

Module 6   **Linear Momentum**

4. Choosing a convenient frame of reference, express the velocity components in terms of the initial and final speeds and the angles of the paths with respect to the horizontal.

$V_{orange-initial-x} =$ $V_{orange-final-x} =$ $V_{orange-final-y} =$

$V_{orange-initial-y} =$ $V_{yellow-final-x} =$ $V_{yellow-final-y} =$

Solving the problem

5. Since there is no initial momentum in the y direction, find the relationship between the final speeds by solving the momentum equation for the y component.

6. Solve the x component momentum equation to find the relation between the initial and final speeds.

PROBLEM 40   (SCREEN 6.7)

Center of Mass

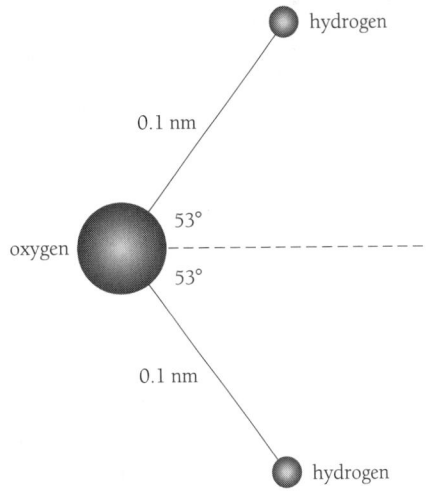

Problem Description

A water molecule consists of an oxygen atom with two hydrogen atoms bound to it. The angle between the two bonds is 106°. If each bond is 0.100 nm long, where is the center of mass of the molecule?

Before we begin...   The mass, in grams, of a single atom is given by its mass in AMU's divided by Avogadro's number. The mass of a single atom of H and O are 1.67×10^{-24} g and 2.66×10^{-23} g, respectively.

1. Assuming that the oxygen atom in the above drawing is located at the origin and the dashed line is along the x axis, is there any symmetry that will simplify the problem?

2. What is the y component of the center of mass (y_{CM})?

Solving the problem   3. Assuming that, because of symmetry, $y_{CM} = 0$, find the x component of each hydrogen atom's position.

4. Write the equation for the location of the x_{CM} for a set of discrete particles, and substitute the values of mass and position for this problem.

MODULE 7

Rotational Mechanics

INTRODUCTION

To describe the motion of a system, we must go beyond translational motion and also consider rotational motion. This module shows the analogies between translational and rotational concepts of position, velocity, and acceleration. In addition, rotational counterparts for mass and force are introduced.

Rotational work, energy, and power are discussed, as are angular momentum and its conservation.

DEFINITIONS

Angular velocity, ω. The rate of change of angular position θ with respect to time. The direction of the angular velocity vector is determined by using the right-hand rule.

Angular acceleration, α. The rate of change of the angular velocity vector with respect to time.

Moment of inertia. The rotational counterpart of mass, represented by the mass multiplied by the square of the distance from the axis of rotation.

Torque, τ. The rotational counterpart of force. Torque depends upon the applied force and the position of the applied force with respect to the point of rotation.

Angular momentum. The rotational counterpart to linear momentum. Just as linear momentum is equal to the product of mass and velocity, angular momentum is computed as the product of moment of inertia and the angular velocity vector.

Law of conservation of angular momentum. In the absence of a net external torque, the angular momentum of a system remains constant.

USEFUL EQUATIONS

In solving rotational kinematic problems, the equations of motion mirror the equations from translational kinematics for constant acceleration. In particular, for constant angular acceleration α,

$$\theta = \theta_0 + \omega_0 t + (1/2)\alpha t^2$$

$$\omega = \omega_0 + \alpha t$$

$$\theta = \theta_0 + (1/2)(\omega_0 + \omega)t$$

$$\omega^2 = \omega_0^2 + 2\alpha(\theta - \theta_0)$$

The moment of inertia for a system of discrete particles is

$$I = \sum m_i r_i^2$$

where r is the perpendicular distance of the particle of mass m from the axis of rotation.

The rotational kinetic energy K_R is found from the equation

$$K_R = (1/2)I\omega^2$$

The parallel axis theorem relates the moment of inertia I about an axis to the moment of inertia about the center of mass I_{CM}.

$$I = I_{CM} + Md^2$$

where d is the distance from the center of mass to the axis of rotation and M is the mass.

Torque about an axis of rotation is evaluated by the relation

$$|\vec{\tau}| = |\vec{r}||\vec{F}|\sin\theta$$

where $\vec{F}$ is the applied force, $\vec{r}$ is the displacement vector from the axis to the point where the force is applied, and θ is the angle between the direction of the force and the line running from the axis to the point where the force is applied.

Work and power are computed using the equations

$$\Delta W = \tau \Delta\theta \quad \text{and} \quad P = \vec{\tau} \cdot \vec{\omega}$$

The angular momentum $\vec{L}$ of a body with moment of inertia I and angular velocity $\vec{\omega}$ is found using the equation

$$\vec{L} = I\vec{\omega}$$

Newton's second law for rotational systems is

$$\vec{\tau}_{net} = I\vec{\alpha}$$

which can also be expressed as

$$\vec{\tau}_{net} = \frac{\Delta \vec{L}}{\Delta t}$$

Module 7 **Rotational Mechanics**

PROBLEM 41 (SCREEN 7.2)

Rotational Kinematics

Problem Description

A dentist's drill starts from rest. After 3.20 s of constant angular acceleration, the drill is turning at a rate of 2.51×10^4 revolutions per minute. Find the drill's angular acceleration, and the angle (in radians) that it rotates through during the 3.20-s interval.

Before we begin...

1. Identify the relevant information:

    initial angular velocity of the drill $\omega_0 =$

    final angular velocity of the drill $\omega_f =$

    elapsed time over which the drill accelerates $\Delta t =$

    (*Note*: One of these relevant quantities was given in units of revolutions per minute. It should be converted into the more standard units of radians per second.)

2. While learning to solve rotational-motion problems, remember to take advantage of the analogies between angular and linear motion. What sort of linear-motion problem is most closely analogous to this rotational problem? Can you use this analogy to help devise a strategy for this problem?

Solving the problem

3. Because the angular acceleration is assumed constant, we can use the angular kinematics equations for constant acceleration. Which of these equations would be best suited for relating the given information to the drill's angular acceleration α?

4. Substitute the known values for all variables in the equation you've chosen and calculate the value of α. Be sure that your answer has the appropriate units for angular acceleration.

5. To find the angle $\Delta\theta$ that the drill turns through while accelerating, we can again use one of the constant-acceleration angular kinematics equations. Based on the information you now know, which of these equations is best suited for the task?

6. Substitute known values into this equation and calculate the value of $\Delta\theta$.

7. While angular calculations are best carried out in units of radians, their answers are sometimes easier to understand if given in rotations.
How many times did the drill bit rotate while accelerating to full speed?

PROBLEM 42 (SCREEN 7.2)

Rotational Kinematics

Problem Description
A discus thrower accelerates a discus from rest to a speed of 25.0 m/s by whirling it through 1.25 revolutions around a circular trajectory of radius 1.10 m. Determine the magnitude of the discus's angular acceleration α, assuming that it is constant. How much time does the thrower take to accelerate the discus from rest to its final speed?

Before we begin...

1. Identify the relevant information:

   initial speed of the discus $v_0 =$

   final speed of the discus $v_f =$

   angle through which the discus rotates $\Delta\theta =$

   radius of the discus's circular path $r =$

   (*Note*: One of these relevant quantities was given in units of revolutions. It should be converted into the more standard units of radians.)

2. For a rotational kinematics problem, it would be more useful to know the discus's initial and final angular speed ω (in radians per second), rather than its linear speed v (in meters per second). How are these two quantities related?

3. Find the discus's initial and final angular speeds:

   initial angular speed $\omega_0 =$

   final angular speed $\omega_f =$

Solving the problem

4. Because the angular acceleration is assumed constant, we can use the angular kinematics equations for constant acceleration. Which of these equations would be best suited for relating the given information to the discus's angular acceleration α?

5. Solve the above equation for α. Does the resulting expression have the correct units for angular acceleration?

6. Substitute known values into the expression from step 5 to find the value of α.

7. Which of the constant-acceleration angular kinematics equations is best suited for finding the acceleration time t from the information you now know?

8. Solve this equation for t and verify that the resulting expression has dimensions of time.

9. Substitute known values into the expression from step 8 to find the value of t.

Module 7 **Rotational Mechanics**

PROBLEM 43 (SCREEN 7.3)

Rotational Kinetic Energy

Problem Description

A car is designed to get its energy from a rotating flywheel, a uniform disc of radius 0.80 m and mass 500 kg. Before the trip, the flywheel is attached to an electric motor which sets it rotating at an initial rate of 8000 rev/min. How much kinetic energy is stored in this flywheel? If the car experiences 1000 N of air resistance at cruising speed (and no other resistive forces of any kind), at what rate will the flywheel be rotating after the car has traveled 50 km? What is the maximum distance the car can travel using the energy stored in the flywheel?

Before we begin...

1. Identify the relevant information:

   radius of the flywheel $R =$

   mass of the flywheel $m =$

   initial angular speed of the flywheel $\omega_0 =$

   force of air resistance on the car $f =$

   distance traveled by car $d =$

   (*Note*: One of these relevant quantities is given in units of revolutions per minute. It should be converted into the more standard units of radians per second.)

2. For a rotational mechanics problem, the flywheel's moment of inertia I is more directly relevant than its mass m. The expression for the moment of inertia of a uniform disc or cylinder is $I = \frac{1}{2}mR^2$. Use this expression and the given information to find I.

Core Concepts in College Physics Workbook

Solving the problem

3. What is the general formula for rotational kinetic energy? Use this formula to find the total energy stored in the flywheel.

4. How much work is done on the car by the air resistance f while it travels a distance d? Write an expression for the energy left in the flywheel after traveling this distance.

5. Using your answer to step 4, along with the general expression for the energy of a rotating body, derive an expression for the flywheel's angular speed ω after the car travels a distance d.

6. Substitute known values into the expression from step 5 to find the flywheel's rate of rotation after the car has traveled $d = 50$ km.

When the car has traveled its maximum possible distance d_{max}, the flywheel will have spent all its energy and will no longer be rotating at all; it will have $\omega = 0$.

7. By setting the expression for ω from step 5 equal to zero and then solving for d, derive an expression for the maximum distance d_{max} the car can travel.

8. Examine your equation for d_{max}. Are the units appropriate for a distance? Can you explain why each variable appears in the numerator or in the denominator?

9. Substitute known values into your expression for d_{max} and find the range of the car in kilometers.

PROBLEM 44 (SCREEN 7.5)

Torque

Problem Description

A spherical satellite of uniform density, with mass 3.0 × 10⁵ kg and radius 2.0 m, floats at rest in deep space. Four rocket engines are arranged around the satellite's equator as shown at right. They fire simultaneously, each exerting a force of 2500 N. After 4 minutes, the engines are turned off. What is the net force on the satellite while the engines are firing? The net torque? At the end of four minutes, how quickly is the satellite spinning? How many times does it rotate during this time period?

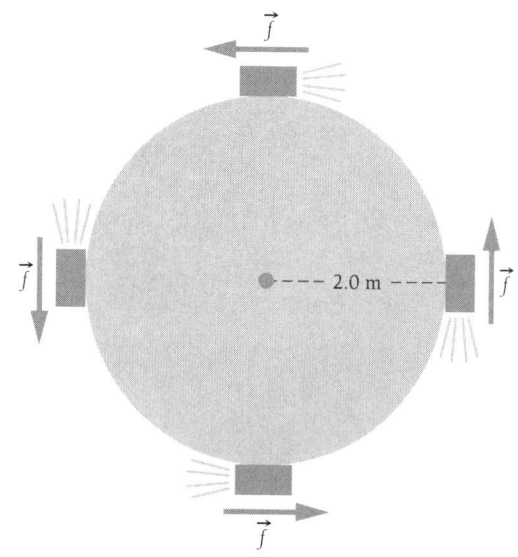

Before we begin...

1. Identify the relevant information:

    radius of the satellite $R =$

    mass of the satellite $m =$

    initial angular speed of the satellite $\omega_0 =$

    force exerted by each rocket engine $f =$

    time for which the engines fire $t =$

2. For a rotational mechanics problem, the satellite's moment of inertia I is more directly relevant than its mass m. The expression for the moment of inertia of a uniform sphere is $I = 2/5\,mR^2$? Use this expression and the given information to find I.

Core Concepts in College Physics Workbook

Solving the problem

3. To find the total force on the satellite, add up (as vectors) the forces exerted by each of the four engines. What is the net force? Will the satellite undergo any linear acceleration?

4. How much torque is exerted on the satellite by a single engine? Do the torques exerted by the four engines act in the same direction or in different directions?

5. What is the total torque about the satellite's axis of rotation?

6. Using Newton's second law for rotational motion, find the magnitude of the satellite's angular acceleration α.

7. Because the torque on the satellite remains constant for the four-minute firing time of the engines, we can analyze its motion using the constant-acceleration angular kinematics equations. Which of these equations is best suited to finding the satellite's final angular speed ω? Substitute known values into this equation and calculate ω.

8. Which of the angular kinematics equations is best suited to finding the angle $\Delta\theta$ that the satellite turns through? Substitute known values into this equation and calculate $\Delta\theta$. Don't forget to convert your final answer from radians into rotations.

9. *Bonus*: If you were asked to calculate the satellite's final kinetic energy, can you think of at least three easy ways to do it? Which one of these methods would yield the largest answer for the energy?

PROBLEM 45 (SCREEN 7.6)

Work and Energy in Rotational Motion

Problem Description

A potter's wheel—a thick stone disk of radius 0.50 m and mass 100 kg—is freely rotating at 50 rev/min. The potter can stop the wheel in 6.0 s by pressing a wet rag against the rim. Find the total change in rotational kinetic energy and the net torque to bring the wheel to rest.

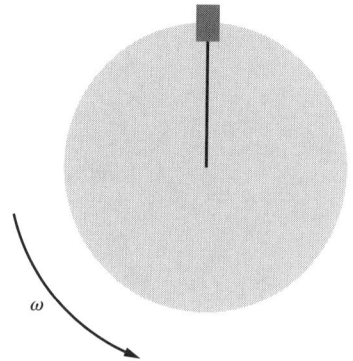

Before we begin...

1. Identify the revelant information:

   mass of the wheel $M =$

   radius of the wheel $R =$

   initial angular speed of the wheel $\omega_i =$

   time over which the wheel decelerates $\Delta t =$

2. The friction from the potter's rag will create a torque about the axis of rotation of the wheel. Will this torque cause the angular speed to increase or decrease? How do you know?

3. What will be the final rotational kinetic energy of the wheel, after the torque has brought the wheel to rest?

Solving the problem

The concepts to be applied to this problem are (1) the work-energy theorem; (2) rotational kinematics; and (3) the definition of torque in terms of applied force and moment arm.

4. To apply the work-energy theorem, we need to recall that the rotational kinetic energy is given by $K = \frac{1}{2}I\omega^2$. Using the equation $I = \frac{1}{2}MR^2$ for a solid disk, calculate the moment of inertia of the wheel.

Core Concepts in College Physics Workbook

5. Next, convert the angular speed, given in rev/min, to radians/second. (Why is this important?)

6. Compute the initial rotational kinetic energy and the change in kinetic energy during the problem.

7. Considering that the work done by the net torque is equal to the change in kinetic energy, write the formula for work done by a constant net torque and set it equal to the change in kinetic energy.

8. Find $\Delta\theta$. Because the angle, $\Delta\theta$, through which the torque was applied was not given, you need to apply the rotational kinematic equations with the given information.

9. Now you can evaluate and solve for the torque.

10. Given the torque and the radius of the wheel, find the magnitude of the frictional force F exerted by the wet rag.

Module 7 **Rotational Mechanics**

PROBLEM 46   (SCREEN 7.7)

Rolling Motion

Problem Description
A uniform solid disk and a uniform hoop are placed side by side at the top of an incline of height h. If they are released from rest and roll without slipping, determine their speeds when they reach the bottom. Which object reaches the bottom first?

Before we begin...

1. Draw a sketch illustrating the problem.

2. What kind of energy does each object have at the top of the incline?

Solving the problem

3. The law of conservation of energy tells us that the gravitational potential energy possessed by the object as it is released to roll down the incline will be converted to kinetic energy as it rolls. Since the object does not slip, no work is done against friction. All of the potential energy lost will result in gained kinetic energy. A rolling object has both rotational kinetic energy and translational kinetic energy. Write the expression for each type of kinetic energy.

Core Concepts in College Physics Workbook

4. Write the law of conservation of energy as it applies to this problem.

5. Solve the resulting equation for the speed at the bottom by recalling the relationship between angular speed and translational speed for a rotating object.

The two moments of inertia for the objects are

$$I_{disk} = \frac{1}{2}MR^2 \qquad I_{hoop} = MR^2$$

6. Substitute the known quantities into the equation. Do this separately for the disk and the hoop. Which has the greater speed at the bottom?

7. The object with the greater speed at the bottom will have reached the bottom first. Why?

PROBLEM 47 (SCREEN 7.9)

Conservation of Angular Momentum

Problem Description
A merry-go-round of radius $R = 2.0$ m has a moment of inertia $I = 250$ kg·m² and is rotating at 10 rev/min. A 25-kg child, who is initially at rest, steps onto the edge of the merry-go-round. What is the new angular speed of the merry-go-round?

Before we begin...

1. State the law of conservation of angular momentum.

2. How is the angular momentum related to the moment of inertia of a system?

3. Identify the relevant information:

   merry-go-round's moment of inertia $I_{m\text{-}g\text{-}r} =$

   radius of the merry-go-round $R_{m\text{-}g\text{-}r} =$

   merry-go-round's initial angular speed $\omega_{m\text{-}g\text{-}r} =$

   mass of the child $m_c =$

4. What is the equation for computing the moment of inertia of a revolving point mass particle? (The child will be considered a point mass in this problem.)

Solving the problem

Because no net external torque is applied to the system (child plus merry-go-round), the angular momentum of the system will remain unchanged. Further, the total moment of inertia of the system about the point of rotation equals the sum of the individual moments of inertia of the objects in the system.

5. Evaluate the angular momentum of the merry-go-round.

6. After the child jumps onto the merry-go-round, what is the new moment of inertia of the system?

7. Use the information to solve for the new rotational speed after the child comes aboard.

MODULE 8

Simple Harmonic Motion and Waves

INTRODUCTION

Simple harmonic motion (SHM) is the periodic oscillation about a stable equilibrium point. SHM occurs when the net force obeys Hooke's law, i.e., it is a restoring force proportional to the displacement and always directed toward the equilibrum point. A wave, on the other hand, can be thought of as a series of simple harmonic oscillators incrementally "out of phase" with each other.

In this module, we explore both these forms of periodic motion. We learn to describe them mathematically and to develop an understanding of the linear restoring forces that drive them.

DEFINITIONS

Amplitude, A. The maximum displacement from the equilibrium position.

Period of motion, T. The time required for one complete oscillation or cycle.

Angular frequency, ω. A constant related to the period of motion by $\omega = 2\pi/T$.

Phase constant, ϕ. An angular offset that describes where a wave or oscillation is in its cycle at time $t = 0$.

Wavelength, λ. The distance between one peak of a wave and the adjacent peak (or one trough and the adjacent trough).

Wave number, k. A number equal to 2π divided by the wavelength.

USEFUL EQUATIONS

When analyzing the motion as a result of applied forces, use Newton's second law. In SHM, we use Hooke's law:

$$F_{net} = ma = -kx$$

Core Concepts in College Physics Workbook

Position as a function of time for SHM is given by

$$x = A \cos(\omega t + \phi)$$

Velocity as a function of time is

$$v = -\omega A \sin(\omega t + \phi)$$

Acceleration as a function of time is

$$a = -\omega^2 A \cos(\omega t + \phi)$$

Upon substituting these equations into Hooke's law, the angular frequency for the system is given by

$$\omega = \sqrt{\frac{k}{m}} \quad \text{for the spring-mass system}$$

$$\omega = \sqrt{\frac{g}{l}} \quad \text{for the simple pendulum}$$

The period of motion T and the frequency f are related to the angular frequency by

$$T = \frac{2\pi}{\omega} \quad \text{and} \quad f = \frac{\omega}{2\pi}$$

The general wave function for a sinusoidal wave moving to the right is of the form

$$y(x, t) = A \cos(kx - \omega t + \phi)$$

The velocity v_y in the direction of displacement y at any point x, at time t, of such a wave is given by

$$v_y(x, t) = \omega A \sin(kx - \omega t + \phi)$$

The acceleration a_y in the direction of displacement is given by

$$a_y(x, t) = -\omega^2 A \cos(kx - \omega t + \phi)$$

The wavelength λ is related to the wave number k by

$$\lambda = \frac{2\pi}{k}$$

The speed of a wave v is related to the wavelength and frequency by

$$v = f\lambda$$

In terms of the wave number and angular frequency, the wave speed is

$$v = \left(\frac{\omega}{2\pi}\right)\left(\frac{2\pi}{k}\right) = \frac{\omega}{k}$$

Module 8 **Simple Harmonic Motion and Waves**

PROBLEM 48 (SCREEN 8.3)

Simple Harmonic Motion

Problem Description

A 20-g particle oscillates with a frequency of 3.0 Hz and an amplitude of 5.0 cm. Through what total distance does the particle move during one cycle of its motion? What is its maximum speed and where does it occur? Find the maximum acceleration of the particle. Where in the motion does the maximum acceleration occur?

Before we begin...

1. Does the mass of the particle matter?

2. Identify the given information:

    $f =$ $A =$

3. Draw a sketch of the position as a function of time for the particle.

Core Concepts in College Physics Workbook

Solving the problem

4. To evaluate the total distance traveled, examine the graph you just drew. What is the distance between the equilibrium point and the maximum displacement? This distance is traveled four times. Verify this on your sketch.

5. If the equation for the displacement as a function of time is $y(t) = A \cos(2\pi ft + \phi)$, write an expression for $v(t)$. What is the maximum value that this expression can have?

6. Use the above information and the general expression for acceleration in SHM to write an expression for $a(t)$. What is the maximum value that this expression can have?

PROBLEM 49 (SCREEN 8.6)

Physical Nature of Waves

Problem Description

A transverse wave on a string is described by $y = (0.12 \text{ m}) \sin \pi[(x/8) + 4t]$. Determine the transverse speed and acceleration of the string at $t = 0.20$ s for the point on the string located at $x = 1.6$ m. What are the wavelength, period, and speed of propagation of this wave?

Before we begin...

1. Write the general expression for a traveling wave on a string.

2. What are the expressions for the transverse speed and acceleration for such a wave?

Solving the problem

3. Compare the general expression for a traveling wave to the equation given in this problem. What are the values of A, w, k, and ϕ?

   A =   w =   k = ϕ =

   Identify the requested values by inspection.

   λ =   T =   v =

4. Substitute the values into the expressions for transverse velocity and transverse acceleration and solve for $x = 1.6$ m and $t = 0.20$ s.

Core Concepts in College Physics Workbook

PROBLEM 50   (SCREEN 8.6)

Frequency, Wavelength, and Wave Speed

Problem Description
A bat can detect a small object, such as an insect, whose size is approximately equal to one wavelength of the sound the bat makes. If bats emit a chirp at a frequency of 60.0 kHz, and if the speed of sound in air is 340 m/s, what is the smallest insect a bat can detect?

Before we begin...

1. What is the relationship between wavelength and frequency?

2. Identify the given information:
   frequency $f =$
   wave speed $v =$

Solving the problem

3. Solve for the wavelength in terms of frequency and wave speed.

4. Substitute the known values to get a value for the wavelength.

Module 8   **Simple Harmonic Motion and Waves**

PROBLEM 51 (SCREEN 8.6)

Frequency, Wavelength, and Wave Speed

Problem Description
A piano emits frequencies that range from a low of about 28 Hz to a high of about 4200 Hz. Find the range of wavelengths spanned by this instrument. The speed of sound in air is approximately 343 m/s.

Before we begin...

1. What is the relationship between wavelength and frequency?

2. Identify the given values:

    lowest frequency $f_{low} =$

    highest frequency $f_{high} =$

    wave speed $v =$

Solving the problem

3. Solve for the wavelength in terms of frequency and wave speed.

4. Substitute the known values to get a value for the high and low wavelengths.

Core Concepts in College Physics Workbook

PROBLEM 52 (SCREEN 8.8)

Mathematical Nature of Waves

Problem Description
A sinusoidal wave traveling in the −x direction (to the left) has an amplitude of 20.0 cm, a wavelength of 35.0 cm, and a frequency of 12.0 Hz. The displacement of the wave at $t = 0$, $x = 0$ is $y = -3.00$ cm, and the wave has a positive velocity here. Sketch the wave at $t = 0$. Find the angular wave number, period, angular frequency, and phase velocity of the wave. Write an expression for the wave function $y(x, t)$.

Before we begin...

1. Identify the given information:

   $A =$ $\lambda =$ $f =$ $y(0, 0) =$

2. Sketch the wave at $t = 0$. In making this sketch, you are taking a snapshot of the wave at $t = 0$. You are actually plotting y as a function of x. What do you have to know in order to make this plot?

3. Write the general expression for the wave traveling to the left as a function of position and time.

4. How will knowing the value of y at $t = 0$ and $x = 0$ help you find the phase constant?

(*continued on next page* . . .)

Module 8 **Simple Harmonic Motion and Waves**

Solving the problem

5. Use the values of wavelength and frequency to calculate the wave number, angular frequency, period, and phase velocity (also known as the wave speed).

6. Evaluate the phase constant by substituting $x = 0$ and $y = 0$ into the traveling wave equation and solving for ϕ. (*Hint*: Use the sign of the velocity to resolve any ambiguity in ϕ.)

7. Write the equation for this particular traveling wave using the general form and the computed values.

PROBLEM 53 (SCREEN 8.9)

Mathematical Nature of Waves

Problem Description
A harmonic wave train is described by $y = (0.25 \text{ m}) \sin(0.30x - 40t)$, where x and y are in meters and t is in seconds. Determine for this wave the amplitude, angular frequency, wave number, wavelength, wave speed, and the direction of the propagation.

Before we begin...

1. Notice that in this problem there does not appear to be a phase angle. The equation is expressed in terms of a sine function rather than a cosine function. This will not affect our results. Write the general expression for the wave function of a traveling wave.

2. According to the general expression, is this wave traveling in the $+x$ or $-x$ direction?

Solving the problem

3. Do a term-by-term comparison of the specific equation and the general equation to identify the following:

   $A =$ \qquad $k =$ \qquad $\omega =$

4. Using the general wave equation, $v = f\lambda$, and the relationships between k and λ and between ω and v, find the wavelength and wave speed.

5. To determine the direction of the propagation, check the sign before the t term in the equation. If it is positive, the motion is to the left. If the sign is negative, the motion is to the right ($+x$ direction).

Module 8 **Simple Harmonic Motion and Waves**

PROBLEM 54   (SCREEN 8.10)

Hooke's Law and the Equation of Motion
Problem Description
A 1.0-kg mass attached to a spring of force constant 25 N/m oscillates on a horizontal, frictionless track. At $t = 0$, the mass is released from rest at $x = -3.0$ cm. (That is, the spring is compressed by 3.0 cm.) Find the period of its motion, the maximum values of its speed and acceleration, and the displacement, velocity, and acceleration as functions of time.

Before we begin...

1. Identify the given variables and terms:

    $m =$ $k =$ $A =$

2. What are the three subproblems to this problem?

Solving the problem

3. What is the relation between the given information and the period of a spring-mass system? Use this to evaluate T.

4. To compute the maximum speed, do the following. In finding the maximum speed of the mass, employ the law of conservation of energy, and then calculate the work done in compressing the spring from equilibrium ($W = (1/2)kA^2$). Since the total energy of the spring-mass system before the work is done was zero, the total energy after compressing the spring will be equal to the work done upon the spring by the external agent.

The maximum speed will occur when all of the energy of the system is kinetic (potential energy will be zero).

5. Finding the maximum acceleration requires recalling that the force exerted by the spring on the mass obeys Hooke's law ($\vec{F} = -k\vec{x}$). Accordingly, if Newton's second law of motion ($\vec{F}_{net} = m\vec{a}$) is employed, the acceleration will be a maximum when the displacement is a maximum. The direction of the acceleration will not be considered. We are asked to find its maximum value only.

(continued on next page . . .)

6. To identify the forms expressing displacement, velocity, and acceleration as functions of time, recall that the general position as a function of time for a system in SHM is

$$x = A \cos(\omega t + \phi)$$

and that

$$v(t) = -\omega A \sin(\omega t + \phi)$$
$$a(t) = -\omega^2 A \cos(\omega t + \phi)$$

Substituting the given value of A and the computed value of ω into the equation for position as a function of time, evaluate the initial conditions to find ϕ. Substitute A, ω, and ϕ into the equation for velocity and acceleration.

PROBLEM 55 (SCREEN 8.11)

SHM and Waves in the Real World

Problem Description
A simple pendulum has a length of 3.00 m. Determine the change in its period if it is taken from a point where $g = 9.80$ m/s² to an elevation where the free-fall acceleration decreases to 9.79 m/s².

Before we begin...

1. For small amplitudes, what is the relation between the period of motion of a simple pendulum, the pendulum's length, and the acceleration due to gravity?

2. How do you calculate the change in a quantity as measured under two different conditions?

Solving the problem

3. Use the equation relating the period of motion to the length of a simple pendulum and the acceleration due to gravity to calculate the periods for each of the given acceleration values.

4. Subtract the value computed for $g = 9.80$ m/s² from the value computed when $g = 9.79$ m/s².

Module 8 **Simple Harmonic Motion and Waves**

PROBLEM 56 (SCREEN 8.14)

Wave Speed

Problem Description

Two points, A and B, on the Earth are at the same longitude and 60.0° apart in latitude. An earthquake at point A sends two waves toward B. A transverse wave travels along the surface of the Earth at 4.50 km/s, and a longitudinal wave travels through the body of the Earth at 7.80 km/s. Which wave arrives at B first? What is the time difference between the arrivals of the two waves at B? Take the radius of the Earth to be 6.37×10^6 m.

Before we begin...

1. Is there any conclusion about relative arrival times that can be made even before making any calculations?

2. What must be determined first to calculate the time difference?

3. Draw a sketch of the Earth, indicating points A and B. Draw a line along the path of the longitudinal wave. It might also be helpful to draw lines connecting the center of the Earth with points A and B.

Core Concepts in College Physics Workbook

4. Identify the given information:

   radius of the Earth $R =$

   speed of transverse wave $v_t =$

   speed of longitudinal wave $v_l =$

Solving the problem

5. What are the expressions for each of the distances traveled?

6. Using the expressions for distance and the speed of each wave, write equations for each travel time.

7. Substitute the known values into each expression and calculate each time. The difference between these times is the time delay between the first arrival of each kind of wave.

MODULE 9

Wave Behavior

INTRODUCTION

In Module 8, *Simple Harmonic Motion and Waves*, we introduced wave motion for traveling waves and defined the concepts of wavelength, frequency, and wave speed. In this module, we investigate how boundary conditions and other waves affect the overall propogation of a wave through a medium, as well as how the energy and power of that wave are affected by such conditions.

DEFINITIONS

Boundary. The interface between any two mediums. Waves at a boundary can be reflected and/or transmitted.

Superposition principle. When two waves combine, they pass straight through without interruption or distortion. At a position x, the total disturbance is the sum of the individual disturbances:

$$y(x, t) = y_1(x, t) + y_2(x, t)$$

Interference. The addition or subtraction of the amplitudes of two waves located at the same position and time yields an amplitude that is constructive if the two are in phase and destructive if the two are 180° out of phase.

Standing wave. A wave in a confined region that "oscillates in place" rather than moving across space. Mathematically, these wave patterns always result from interference between two or more moving waves.

Harmonics. Stable modes of vibration corresponding to particular frequencies. The lowest allowable frequency is called the first harmonic.

Node. A point of zero amplitude in a standing wave, created by ongoing destructive interference.

Core Concepts in College Physics Workbook

Antinode. A position of relative maximum amplitude in a standing wave created by ongoing constructive interference.

Resonance. When energy is added to a system at the system's natural frequency, the amplitude of oscillations is maximized.

USEFUL EQUATIONS

The speed of a one-dimensional mechanical wave such as a pulse in a stretched string is computed by

$$v = \sqrt{\frac{F_t}{\mu}}$$

where F_t is the tension in the string and μ is the linear mass density of the string.

The energy E and the power P associated with a wave are calculated by the equations

$$E = (1/2)\mu\omega^2 A^2 L \quad \text{and} \quad P = (1/2)\mu\omega^2 A^2 v$$

where ω is the angular frequency of the wave, A is the wave amplitude, L is the length of the string, and v is the speed of the wave.

For standing waves, the relationships between the allowable frequencies of vibration and the length of the string are given by

$$f_n = \frac{nv}{2L} \quad n = 1, 2, 3... \quad \text{(wave fixed at both ends or free at both ends)}$$

$$\lambda_n = \frac{2L}{n}$$

$$f_n = \frac{nv}{4L} \quad n = 1, 3, 5... \quad \text{(wave fixed at one end and free at the other end)}$$

$$\lambda_n = \frac{4L}{n}$$

Module 9 **Wave Behavior**

PROBLEM 57 (SCREEN 9.2)

Speed of a Wave in a Medium

Problem Description

A light string of mass per unit length 8.00 g/m has its ends tied to two walls separated by a distance equal to 3/4 the length L of the string. A mass m is suspended from the center of the string, putting a tension in the string. Find an expression for the transverse wave speed in the string as a function of the hanging mass. How much mass should be suspended from the string to have a wave speed of 60.0 m/s?

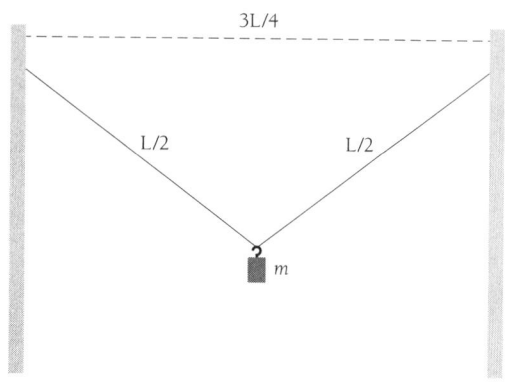

Before we begin...

1. Identify the given information:

    linear mass density of the string $\mu =$

    desired wave speed $v =$

2. Draw the free-body diagram for the mass.

Solving the problem

The relation between the transverse wave speed and the string through which it is traveling is

$$v = \sqrt{\frac{F_t}{\mu}}$$

where F_t is the tension and μ is the linear mass density.

We already know μ, but not F_t. Thus the main subproblem we must solve is finding the tension F_t.

3. The tension in each section of the string is the same. This ensures that the x components of the forces sum to zero. What do the y components of the tensions have to equal to ensure equilibrium?

4. How can the angle θ that the string makes with respect to the horizontal be computed?

5. The string and the mass are in equilibrium, requiring that the vector sum of the forces be zero. By symmetry, the x components of the tension will be equal in magnitude to each other. Apply the condition of equilibrium to the y components and solve for the tension T.

6. Substitute this expression for F_t into the equation for the speed v of waves on a string to find an expression for v in terms of m, μ, and θ.

7. Solve the expression from step 6 for m. In order for the wave speed v to be 60 m/s, what does m have to be?

PROBLEM 58 (SCREEN 9.5)

Energy and Power in Waves

Problem Description

Transverse waves are being generated on a rope under constant tension. By what factor is the required generating power increased or decreased if the length of the rope is doubled and the angular frequency remains constant? The power is changed by what factor when the amplitude is doubled and the angular frequency is halved? When both the wavelength and amplitude are doubled, what happens to the power? What happens to the power when both the length of the rope and the wavelength are halved?

Before we begin...

1. Write the expression for the power delivered by a transverse wave.

2. Does the length of the rope affect the power?

Solving the problem

3. If the length changes, it does not affect the power. Isolate the dependency for the various combinations of variables described in the four questions asked in the description. Answer the questions.

PROBLEM 59 (SCREEN 9.7)

Superposition and Interference

Problem Description

Two waves are traveling in the same direction along a stretched string. Both waves have the same frequency and wavelength. Each has an amplitude of 4.0 cm, and they are 90° out of phase. Find the amplitude of the resultant wave.

Before we begin...

1. Write the wave function of each of the two waves.

2. State the superposition principle.

Solving the problem

3. Recall the trigonometric identity

$$\sin(a) + \sin(b) = 2 \sin((a+b)/2) \cos(a-b)/2$$

Let $a = (kx - \omega t)$ and $b = (kx - \omega t - \phi)$. In this problem, $\phi = 90°$. Substitute and simplify the equation.

PROBLEM 60 (SCREEN 9.8)
Standing Waves

Problem Description
Two waves on a string given by
$y_1(x, t) = A \sin(kx - \omega t)$ and
$y_2(x, t) = A \sin(2kx + \omega t)$ interfere.
Determine all x values where there are stationary nodes. At points which are not nodes, the string will pass through the equilibrium position $y(x, t) = 0$. Find all values of x where $y(x, t) = 0$.

Before we begin...

1. What are node positions?

2. According to the superposition principle, what must be true for a point to be a node?

Solving the problem

3. Use the superposition principle to find the wave function of the combined waves. *Hint:* Use the trigonometric relation

$$\sin \alpha + \sin \beta = 2 \sin\left(\frac{\alpha + \beta}{2}\right) \cos\left(\frac{\alpha - \beta}{2}\right)$$

4. Stationary nodes occur as a result of a term that is independent of time t equaling zero. Identify the term in the combined equation, set it equal to zero, and solve for the values of x that satisfy the equation.

5. Time dependent nodes arise from a term that has both position x and time t dependency. Identify such a term in the equation and solve it by setting it equal to zero.

PROBLEM 61   (SCREEN 9.9)

Standing Waves—Wave Fixed at Both Ends

Problem Description
A 2.0-m long wire having a mass of 0.10 kg is fixed at both ends. The tension in the wire is maintained at 20 N. What are the frequencies of the first three allowed modes of vibration?

Before we begin...

1. Identify the relevant quantities:

   mass of wire $M =$

   length of wire $L =$

   tension in wire $F_t =$

2. What is the relationship for the allowable wavelengths for a wave fixed at both ends?

3. How can the speed of the wave be calculated for a wire under tension?

Solving the problem

4. Compute the mass density μ for the wire and use this result to find the speed of the wave.

Core Concepts in College Physics Workbook

5. Combine the relation $v = f\lambda$ with your answer to step 2 to find an expression for the allowable frequencies f_n.

6. Substitute known values into this equation to find f_1, f_2, and f_3.

PROBLEM 62 (SCREEN 9.9)

Standing Waves—Wave with One Fixed End and One Free End

Problem Description

A student uses an audio oscillator of adjustable frequency to measure the depth of a water well. Two successive resonant frequencies are heard at 52.0 Hz and 60.0 Hz. How far beneath the top of the well is the water level?

Before we begin...

Sound waves are longitudinal—that is, they cause the air molecules to vibrate back and forth along the direction that the wave is moving. The air adjacent to the water cannot vibrate in this way—it would have to pass through the water's surface—so the wave must have a node at water level. At the top of the well is an antinode, where the air can vibrate freely. Thus, the wave is similar to a wave on a string with one end fixed and one end free.

1. What is the relationship between the length of the system and allowable frequencies for this type of system?

2. Identify the given information:

    $f_a =$ $f_b =$

Core Concepts in College Physics Workbook

Solving the problem

3. Write the expression for the allowable frequency with $a = n$ and for the allowable frequency with b = next allowable frequency.

4. Solve the equations simultaneously by setting the difference in frequencies $f_b - f_a$ equal to the difference in their computational formulas. This will eliminate n from the equation. We now have change in frequency as a function of wavespeed and depth of the well.

5. Use the speed $v = 344$ m/s (speed of sound at standard temperature and pressure). Solve for the depth to the water surface.

MODULE 10

Thermodynamics

INTRODUCTION

The laws of thermodynamics allow us to express the relationship between thermal energy transfer, work, and internal energy. These laws also provide limits on the efficiency of thermal processes.

The classical view of thermodynamics allows us to analyze the effects of thermal energy on the gross properties of matter. Our study centers around what happens to the system. We use the statistical mechanical approach to give insight into the thermal processes at the molecular level and then relate the properties to measurable quantities such as pressure and temperature.

DEFINITIONS

Internal energy. The collective term for all forms of energy internal to a substance (not influenced by the overall translation or rotation of the body as a whole). Chemical energy, nuclear energy, and thermal energy are internal energy types.

Thermal energy. The total internal mechanical energy of the molecules of a substance.

Thermal equilibrium. The process that occurs when two bodies are placed in thermal contact and no net heat flows between them. (Heat is another name for thermal energy transfer.)

Temperature. If two bodies placed in thermal contact remain in thermal equilibrium, then they are said to have the same temperature. If they do not, then heat will flow from the body with higher temperature to the one with lower temperature. For most types of systems, temperature is proportional to the average kinetic energy of the molecules.

Heat capacity. A measure of the amount of thermal energy that is required to raise an object's temperature by a specific amount.

Absolute zero. The lowest possible temperature. At this temperature, molecules of a substance have essentially zero thermal energy. Absolute zero corresponds to -273.15 °C.

Zeroth law of thermodynamics. Two objects that are independently in thermal equilibrium with a third object are in thermal equilibrium with each other.

First law of thermodynamics. The first law of thermodynamics is a generalization of the law of conservation of energy to include thermal energy. Any thermal energy absorbed by a system increases the system's internal energy or goes into work done by the system, or both.

Adiabatic process. A process that occurs when there is no net thermal energy transfer.

Isothermal process. A process that occurs when there is no change in temperature.

Second Law of Thermodynamics. In a closed system, the total entropy either increases or stays the same.

Entropy. A measure of the disorder of a system.

USEFUL EQUATIONS

The heat capacity C of a substance is calculated by the equation

$$C = \frac{Q}{\Delta T}$$

where Q is the thermal energy transferred and ΔT is the change in temperature.

The ideal gas law relating pressure, volume, and temperature is described by the equation

$$PV = Nk_BT$$

where N is the number of gas molecules in the sample and k_B is a constant known as Boltzmann's constant.

The first law of thermodynamics can be expressed as

$$\Delta Q = \Delta W + \Delta U$$

with W representing the work done by the system and U being the internal energy.

Module 10 **Thermodynamics**

The efficiency of a heat engine is computed by

$$e = \frac{\text{work}}{\text{energy absorbed}} = \frac{Q_h - Q_c}{Q_h}$$

where Q_h is the heat absorbed from the hot reservoir and Q_c is the heat deposited to the cold reservoir.

For a Carnot engine, the efficiency is

$$e = \frac{T_h - T_c}{T_h}$$

with the temperatures being measured on the absolute temperature scale.

The coefficient of performance (COP) of a Carnot heat pump is

$$\text{COP} = \frac{Q_h}{W} = \frac{T_h}{T_h - T_c}$$

The entropy of a system is computed by using the equation

$$\Delta S = \frac{\Delta Q_r}{T}$$

where ΔQ_r is the change in thermal energy along a reversible path.

PROBLEM 63 (SCREEN 10.3)

Basic Concepts of Thermodynamics

Problem Description

A perfectly insulated calorimeter contains 500 ml of water at 30°C and 25 g of ice at 0°C. Determine the final temperature of the system.

Before we begin...

1. The system is insulated and isolated from its surroundings. It will come to thermal equilibrium. What is meant by thermal equilibrium?

   The concept of heat capacity C is discussed on the CD-ROM. Associated with the concept is another concept called specific heat c. Specific heat is the amount of heat per unit mass required to change the temperature of a substance by a given amount.

   $$c = \frac{Q}{m\Delta T}$$

   For a substance to change phase, an additional heat transfer is required. L_f is the amount of heat transferred per unit mass as the substance melts or freezes, and L_v is the heat transferred per unit mass as it vaporizes or condenses.

   $$Q = \pm mL_f \quad \text{or} \quad Q = \pm mL_v$$

2. Identify the given information:

   volume of water $V_w =$

   initial temperature of water $T_{1w} =$

   mass of ice $m_{ice} =$

   initial temperature of ice $T_{1\text{-}ice} =$

(continued on next page . . .)

The constant values required for this problem are

specific heat of water $c = 4186$ J/kg C°

heat of freezing for water/ice $L_{f\text{-ice}} = 333$ kJ/kg

melting point of ice $T_{mp\text{-ice}} = 0°C$

The density of water is taken to be 1 g/cm³.

Solving the problem

3. We must pay attention to the units in the given information and convert to consistent units throughout the problem. Determine the mass, in kilograms, of the water, using 1.00 ml = 1.00 cm³.

4. In this problem, the net heat exchange will equal zero and the system will reach thermal equilibrium. The water will lose thermal energy, and the ice will use thermal energy to undergo a change of phase. If additional thermal energy is available after all the ice is melted, the ice water will be warmed. Calculate the thermal energy that would be released by lowering the water to the freezing point 0° C.

5. Determine the thermal energy required to melt all of the ice.

6. Solve for the final temperature in this case by applying the law of conservation of energy.

Core Concepts in College Physics Workbook

PROBLEM 64 (SCREEN 10.5)

The Ideal Gas

Problem Description

An ideal gas is held in a container at constant volume. Initially, its temperature is 10.0° C and its pressure is 2.50 atm. What is its pressure when its temperature is 80.0° C?

Before we begin...

1. State the ideal gas law.

2. Identify the given information:

    initial temperature $T_1 =$

    final temperature $T_2 =$

    initial pressure $P_1 =$

Solving the problem

3. In solving problems using the ideal gas law, the temperature must be expressed in terms of absolute (Kelvin) temperature. Convert the temperatures T_1 and T_2 to kelvins.

4. The ideal gas law can be rewritten as

$$\frac{P}{T} = \frac{Nk_B}{V}$$

Since neither N (the number of gas molecules) nor V (the volume of the gas) change, the quantity P/T must also remain constant in this problem. Use this ratio to solve for the new pressure P_2.

Module 10 **Thermodynamics**

PROBLEM 65   (SCREEN 10.6)

The First Law of Thermodynamics

Problem Description

One mole of gas is initially at a pressure of 2.0 atm and a volume of 0.30 L and has an internal energy equal to 91 J. In its final state, the gas is at a pressure of 1.5 atm and a volume of 0.80 L, and its internal energy equals 182 J. For the paths *IAF*, *IBF*, and *IF* in the figure, calculate the work done by the gas and the net heat transferred to the gas in the process.

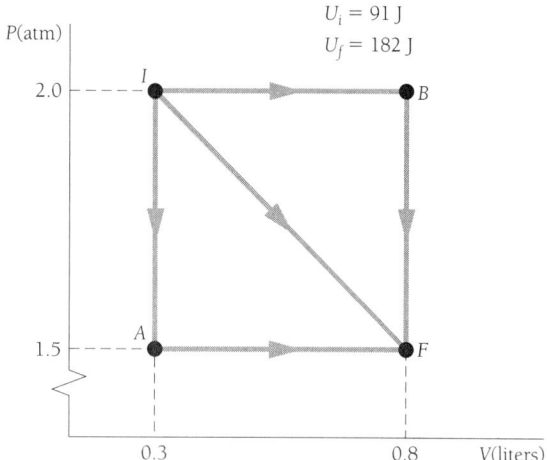

Before we begin...

1. Express the first law of thermodynamics as it applies to the gas, in terms of the change in internal energy, the heat transferred to the gas, and the work done by the gas.

2. In terms of a pressure-volume (*PV*) diagram, what is the work done by a gas in an expansion from some initial state to some final state?

3. What is the expression for the work done at constant pressure?

4. What is the work done at constant volume when the pressure changes?

Core Concepts in College Physics Workbook

5. Identify the given information:

   initial pressure of the gas $P_i =$

   initial volume of the gas $V_i =$

   initial internal energy $U_i =$

   final pressure of the gas $P_f =$

   final volume of the gas $V_f =$

   final internal energy $U_f =$

Solving the problem

6. Break each path *IAF* and *IBF* into two subpaths, one of constant volume, the other of constant pressure. Express the work done on each path in terms of the work done on its subpaths.

7. Use geometrical relationships to express the work on path *IF* in terms of the work done on the other two paths.

8. Using the equation which expresses the first law, solve for the heat which must be transferred to the gas along each path.

9. Substitute the known values in the expressions for work and heat to get numerical values.

PROBLEM 66 (SCREEN 10.9)

Carnot Engines

Problem Description

A given Carnot engine has a power output of 150 kW. The engine operates between two reservoirs at 20° C and 500° C. How much thermal energy is absorbed per hour? How much thermal energy is lost per hour?

Before we begin...

1. Identify the given information:

   temperature of the cold reservoir $T_c =$

   temperature of the hot reservoir $T_h =$

   power output $P =$

2. What is the definition of power P?

3. What does the efficiency of a heat engine measure? Express your answer in terms of work done and energy required.

4. What is the equation for the efficiency of a Carnot engine in terms of the temperatures of the hot and cold reservoirs?

Solving the problem

5. The efficiency of a Carnot engine can be expressed as a function of its absolute temperatures T_c and T_h. Convert the temperatures to kelvins and solve for the efficiency of the engine.

6. Use the definition of power and the given information to find the rate at which work is being done by the engine.

7. Use the definition of efficiency and the rate at which work is done to calculate the thermal energy that is absorbed per hour.

8. Using the definition of work done by a heat engine, compute Q_c, the energy that is lost to the cold reservoir.

PROBLEM 67 (SCREEN 10.9)

Carnot Engines—The Heat Pump

Problem Description
What is the maximum coefficient of performance (COP) of a heat pump that brings heat from outdoors at −3° C into a 22° C house?

Before we begin...

1. The maximum coefficient of performance for a heat pump is its Carnot COP. How is this computed?

2. Identify the given information:

   temperature of hot reservoir T_h =

   temperature of cold reservoir T_c =

Solving the problem

3. The heat pump does work W, which is also available to heat the house. This is already factored into the Carnot COP. Convert the temperatures T_h and T_c to the Kelvin scale.

4. Evaluate the Carnot COP.

Core Concepts in College Physics Workbook

PROBLEM 68   (SCREEN 10.10)

Entropy

Problem Description

What is the change in entropy of 1.0 kg of water at 100° C as it changes to steam at 100° C? A freezer is used to freeze 1.0 L of water completely into ice. The water and the freezer remain at a constant temperature of $T = 0°$ C. Determine the change in the entropy of the water and the change in the entropy of the freezer.

Before we begin...

1. Write the expression for the change in entropy of a system.

2. In general, there could be a temperature change when heat is transferred between a substance and its environment. This would complicate the entropy calculations. For both these processes (water to steam, water to ice), the transition is taking place at a constant temperature, so heat is flowing only due to phase changes. What is the expression for the heat required to change the phase in each of the processes? Which way is the heat flowing in each process?

3. Identify the given information for each part:

   water to steam:

   mass of the water $m_a =$

   temperature $T_a =$

(*continued on next page . . .*)

water to ice:

    volume of the water $V_b =$

    temperature $T_b =$

Other useful information:

    latent heat of freezing for ice $L_f = 3.33 \times 10^5$ J/kg
    latent heat of vaporization for water $L_v = 2.26 \times 10^6$ J/kg

    The density of water is taken to be $\rho = 1$ g/cm³.

Solving the problem

4. In order to use the entropy equation, temperatures must be measured on an absolute scale. Convert the temperatures into the Kelvin scale.

5. The latent heat equations are expressed in terms of the mass of the substance. What is the mass of the water being frozen?

6. For each process, express the change in entropy in terms of the latent heat equations as found in step 2. Use the given information (as expressed in the appropriate units) in the equations derived to calculate the heat required for each of the phase changes. The heat of fusion is used when a substance changes from a solid to a liquid. When the water is being frozen, heat will be flowing out, and so there must be a factor of negative one.

7. What does conservation of energy say about the relationship between the heat flowing out of the water as it freezes and the heat that is absorbed by the freezer? Use this to calculate the change in entropy of the freezer.

MODULE 11

The Electric Field

INTRODUCTION

Electric forces act over a distance between charged objects and can be either attractive or repulsive. It is convenient to think of these forces as being exerted by a vector quantity called the "electric field," which can have a different magnitude and direction at each point in space. A particle's charge measures its tendency to affect and be affected by electric fields.

In this module, we examine electric fields and the related concepts of electric potential energy and electric potential. Gauss's law provides a useful method of computing the electric field in regions of high symmetry.

DEFINITIONS

Electric charge. Charge is an intrinsic feature of every particle. Charge can be either negative or positive. The total charge on an object is the sum of the charges on the particles that make it up. The SI unit of charge is called the coulomb.

Insulator. An electric insulator resists the movement of charge.

Conductor. An electric conductor allows charge to move easily within it.

Coulomb's law. The force between any two stationary charges is directly proportional to the product of the charges and inversely proportional to the square of the distance separating the charges. Opposite-signed charges attract each other while like-signed charges repel.

Electric field lines. Electric field lines point in the same direction as the electric force on a positive charge at every point in space. Field lines begin on positive charges and end on negative charges. No two electric field lines can cross or touch.

$\phi = EA\cos\theta$

$\phi = \dfrac{q_{enc}}{\varepsilon_0}$

Electric flux. The electric flux ϕ_E is proportional to the number of electric field lines that pass through a given area of a surface A.

Gauss's law. The total electric flux through a closed surface is proportional to the charge enclosed by that surface. If a closed surface encloses no net charge, the total electric flux through the surface is zero.

Electric potential energy. Moving a charged particle through an electric field requires that work be done. The negative of this work is the change in electrical potential energy.

Electric potential. The electric potential is defined as the change in electric potential energy per unit charge experienced by a charge when it moves through an electric field. The SI unit of electric potential is the volt.

USEFUL EQUATIONS

Coulomb's law can be expressed in the equation

$$F_{12} = k_e \dfrac{q_1 q_2}{r^2}$$

where k_e is the Coulomb constant (8.99×10^9 N·m/C²), q_1 and q_2 are the charges on the particles, and r is the distance separating the particles. The force is directed along a line joining the particles. It is attractive if the charges are of opposite sign. It is repulsive if the charges have the same sign.

The electric field $\vec{E}$ can be defined by the relationship

$$\vec{E} = \dfrac{\vec{F}}{q_0}$$

with q_0 representing the charge experiencing the force at the particular location.

The electric field a distance r from a single point charge q has magnitude

$$|\vec{E}| = -k\dfrac{q}{r^2}$$

The direction of $\vec{E}$ is directly away from the charge (if q is positive) or toward the charge (if q is negative).

The potential energy U is defined such that the change in potential energy equals negative the work W done by the electric force $q_0 E$.

$$\Delta U = -W$$

Core Concepts in College Physics Workbook

The electric potential difference is computed by dividing the electric potential energy difference by the charge experiencing the field.

$$\Delta V = \frac{\Delta U}{q_0} = \frac{-W}{q_0}$$

The electric potential a distance r from a single point charge q is

$$V = k\frac{q}{r}$$

For the special case of a uniform electric field, when a particle moves between two points, the work is given by

$$W = q_0 E_{\parallel} d$$

and the change in electric potential energy is

$$\Delta U = -W = -q_0 E_{\parallel} d$$

where d is the distance between the points and $E_{\parallel}$ is the component of the electric field parallel to the line joining the points. The electric potential difference in this case is

$$\Delta V = -E_{\parallel} d$$

If the electric potential is known, we can rearrange the above equation and thus can compute the component of the average electric field directed along the line between any two points

$$E_{\parallel} = \frac{-\Delta V}{d}$$

In many cases, the symmetry of the charge distribution, reflected in a symmetric electric field, is such that Gauss's law can be expressed

$$q_{net} = \varepsilon_0 E A$$

where A is the area of a symmetric Gaussian surface, along which E is everywhere perpendicular with a constant magnitude. The constant ε_0 is called the permittivity of free space and is 8.85×10^{-12} C²/N·m². The quantity EA is known as $\boldsymbol{\phi}_E$, the flux of the electric field through the surface A.

Module 11 **The Electric Field**

PROBLEM 69 (SCREEN 11.4)

Coulomb's Law

Problem Description

Four identical point charges of $+10.0\ \mu C$ are located on the corners of a rectangle as shown. The dimensions of the rectangle are $L = 60.0$ cm and $W = 15.0$ cm. Calculate the magnitude and direction of the net electric force exerted on the charge at the lower left corner by the other three charges.

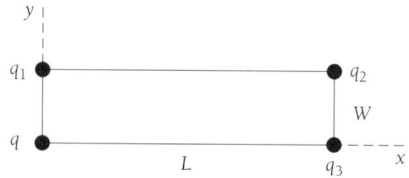

Before we begin...

1. Identify the given information (express length and width in meters):

    charge on each particle $q =$

    length of rectangle $L =$

    width of rectangle $W =$

2. State Coulomb's law for the force between two point charges.

3. How is the direction of the force between two point charges determined?

4. Draw the free-body diagram for the charge whose net force we are to calculate. Use the upper left charge as q_1, upper right q_2, and lower right q_3.

Core Concepts in College Physics Workbook

Solving the problem

5. The electric force $\vec{F}_T$ is equal to the sum of all of the electric forces $\vec{F}_i$ that act upon it. Using Coulomb's law, evaluate the forces $\vec{F}_1$, $\vec{F}_2$, and $\vec{F}_3$.

6. Resolve the forces into their x and y components. Add the x components to yield the total x force. Do the same for the y components.

7. Use the Pythagorean theorem to determine the magnitude of $\vec{F}_T$.

8. Determine the direction of $\vec{F}_T$ by applying the definition of the trigonometry function $\tan\theta$.

Module 11 The Electric Field

PROBLEM 70   (SCREEN 11.5)

The Electric Field and Field Lines

Problem Description

Four point charges are at the corners of a square of side *a* as shown. Determine the magnitude and direction of the electric field at the location of charge *q*. What is the resultant force on *q*?

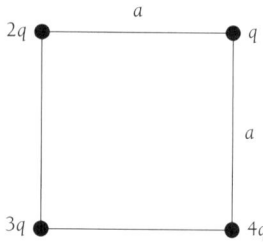

Before we begin...

1. Assume that the charge *q* is located at the origin of a cartesian coordinate system. Sketch the electric fields at the origin due to the other charges.

2. How is the electric field for a point charge calculated?

3. How is the direction of the electric field due to a point charge determined?

4. What is the electric force exerted on a charge *q* by a known field $\vec{E}$?

Solving the problem

5. The electric field $\vec{E}_T$ is equal to the sum of all of the electric fields $\vec{E}_i$ that act upon it. Using the relationship for the electric field due to a point charge, evaluate the fields $\vec{E}_{2q}$, $\vec{E}_{3q}$, and $\vec{E}_{4q}$.

6. Resolve the fields into their *x* and *y* components. Add the *x* components to yield the total *x* field. Do the same for the *y* components.

Core Concepts in College Physics Workbook

7. Use the Pythagorean theorem to determine the magnitude of $\vec{E}_T$.

8. The direction of $\vec{E}_T$ can be determined by applying the definition of the trigonometry function $\tan \theta$.

9. Using the definition of electric fields, find the electric force on charge q by applying $\vec{F} = q\vec{E}$.

PROBLEM 71 (SCREEN 11.6)

Gauss's Law

Problem Description
The total electric flux through a closed surface in the shape of a cylinder is 8.60×10^4 N·m²/C. What is the net charge within the cylinder? How would your answers change if the net flux were -8.60×10^4 N·m²/C?

Before we begin...

1. State Gauss's law for electric fields.

2. Because the total flux is known, does the shape of the closed surface matter when calculating the net charge within the surface? Explain.

3. If the electric flux is negative, what does this tell us about the average relationship between $\vec{E}$ and the surface normal $\vec{n}$?

Solving the problem

4. Use the relationship between total electric flux and the net charge to calculate the charge.

Core Concepts in College Physics Workbook

PROBLEM 72 (SCREEN 11.7)

Examples of the Electric Field

Problem Description

The charge per unit length on a long, straight filament is −90.0 μC/m. Find the electric field 10.0 cm, 20.0 cm, and 100 cm from the filament, where distances are measured perpendicular to the length of the filament.

Before we begin...

1. Assume that the charge per unit length is represented by λ. How much charge would be inside a right cylinder of length l and radius r as shown above?

2. The electric field $\vec{E}$ by symmetry will point radially inward toward the wire, and therefore it will be antiparallel to surface normal $\hat{n}$ through the side of the cylinder. What will the flux be through the two end caps of the cylinder?

Solving the problem

3. Apply Gauss's law to the drawing to derive the electric field at radial distances from a long straight wire.

4. Use this result to evaluate the magnitude of $\vec{E}$ at the given distances.

Module 11 **The Electric Field**

PROBLEM 73   (SCREEN 11.8)

Electric Potential

Problem Description

Two point charges, $q_1 = 5.00$ nC and $q_2 = -3.00$ nC, are separated by a distance of 35.0 cm. What is the electric potential at point A, midway between the two charges? What is the electric field at that same point?

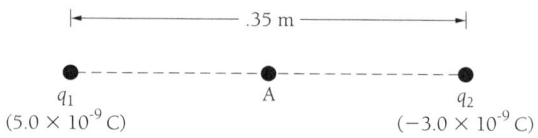

Before we begin...

1. Identify the given information:

   charge $q_1 =$

   charge $q_2 =$

   distance between charges =

   distance from point A to $q_1 =$

   distance from point A to $q_2 =$

2. How do you calculate the electric potential at a given point caused by a collection of point charges?

3. How do you calculate the electric field at a point caused by a collection of point charges?

Core Concepts in College Physics Workbook

Solving the problem

4. Use the given information to compute the electric potential. (If the electric potential is negative, work will have to be done to separate the charges.)

5. Calculate the electric field.

PROBLEM 74 (SCREEN 11.9)

Electric Field and Electric Potential

Problem Description
A pair of oppositely charged, parallel plates are separated by 5.33 mm (a small distance relative to the length and width of the plates). A potential difference of 600 V exists between the plates. What is the magnitude of the electric field between the plates? What is the magnitude of the force on an electron between the plates? Make a sketch showing the direction of the force on the electron. How much work must be done on the electron to move it to the negative plate if it is initially positioned 2.90 mm from the positive plate?

Before we begin...

1. What qualitative deduction about the electric field between the plates can be made because of the relationship of the plates' separation to their dimensions?

2. What is the relationship between a constant electric field $\vec{E}$ and the potential difference ΔV between two points?

Core Concepts in College Physics Workbook

3. What is the relationship between the electric force on a particle and the electric field producing that force?

4. Make a sketch of the plates looking from the "side" so that the separation is visible. Label each plate as being positively or negatively charged; draw a point showing the location of the electron and a vector indicating the direction of the force on the electron. Indicate also where the electric potential is highest. It is helpful to choose a coordinate system with an axis aligned in the direction of motion.

5. How is work defined?

6. Identify the known quantities:

   potential difference between the plates $\Delta V =$

   separation between the plates $s =$

   charge on an electron $q =$

   distance electron is moved $d =$

Solving the problem

7. Substitute the known quantities in the equation from step 2 to find the strength of the electric field.

8. Use this value for the electric field and the value for the electron's charge (-1.6×10^{-19} C) in the equation for force from step 3, to find the magnitude of the force.

9. With the force magnitude solved, the work done in moving the electron can be calculated using the definition of work.

Module 11 The Electric Field

MODULE 12

The Magnetic Field

INTRODUCTION

In Module 11, *The Electric Field*, we discussed the interaction between charged objects in terms of electric fields. In this module, we find that moving charges have another interaction, which we describe in terms of the magnetic field. Magnetic and electric forces are both present when a charged particle moves through a region of space containing magnetic and electric fields. The magnetic field does no work on a moving charged particle because the force exerted is at right angles to the path taken by the particle.

We will explore the links between electric and magnetic fields and discover that changing magnetic fields produce electric fields, and vice versa.

DEFINITIONS

Tesla. The SI unit for the magnetic field $\vec{B}$ is the tesla. One tesla T equals one newton per ampere·meter.

Magnetic force. A charged particle moving relative to an external magnetic field experiences a magnetic force. The force's magnitude is proportional to the charge and speed of the particle, the strength of the magnetic field, and the sine of the angle between the velocity and the field. Thus, the force is zero when the particle is moving parallel to the field, and at a maximum when moving perpendicular to the field. The force's direction, perpendicular to both the velocity and the magnetic field, can be determined using the right-hand rule.

Ampere's law. This law describes the relationship between constant currents and the magnetic fields they produce. Like Gauss's law for electricity, Ampere's law simplifies the calculation of the magnetic field $\vec{B}$ in highly symmetric problems.

Gauss's law for magnetism. This law asserts that the magnetic flux through any closed surface is zero. This is equivalent to the statement that magnetic monopoles do not exist.

Induction. Induction is the process by which changing magnetic flux gives rise to an electomotive force (*emf*) around a closed loop.

Faraday's law of induction is expressed by the equation

$$Emf = -\frac{\Delta \phi_B}{\Delta t}$$

Magnetic flux through a surface is computed by

$$\phi_B = \sum B_\perp \Delta A = \sum B \Delta A \cos \theta$$

For a uniform magnetic field B, the flux through a loop of area A is

$$\phi_B = B_\perp A = BA \cos \theta$$

where θ is the angle between the field and the normal to the plane of the loop.

The magnetic flux can be changed by changing either the magnetic field, the area through which the field points, or the angle between the field and the surface normal.

Lenz's contribution to Faraday's law is the negative sign. The induced *emf* opposes the change in magnetic flux.

USEFUL EQUATIONS

The Lorentz force acting upon a moving charged particle in a region of space with electric and magnetic fields is the vector sum of the electric and magnetic forces.

$$\vec{F}_L = q\vec{E} + \vec{F}_B$$

The magnetic force's magnitude F_B can be calculated by

$$F_B = qvB \sin \theta$$

where θ is the angle between the particle's velocity $\vec{v}$ and the magnetic field $\vec{B}$.

The direction of $\vec{F}_B$ is given by the right-hand rule.

The force $\vec{F}$ on a long straight wire of length L carrying a current $\vec{I}$ in a magnetic field $\vec{B}$ with θ the angle between $\vec{I}$ and $\vec{B}$ is

$$F = ILB \sin \theta$$

Module 12 **The Magnetic Field**

Sometimes the geometry of the current distribution results in a magnetic field constant in magnitude along, and everywhere parallel to, a simple closed path. In such cases, Ampere's law states

$$\mu_0 I_{net} = BS$$

where S is the length of the path and μ_0 is a constant called the permeability of free space. The right-hand rule determines the direction of the associated magnetic field for a current carrying wire.

The general form of Faraday's law allows for induced electric fields. Again, certain symmetric situations give rise to induced electric fields constant in magnitude along, and everywhere parallel to, simple closed paths. Then Faraday's law simplifies to

$$ES = -\frac{\Delta \phi_B}{\Delta t}$$

where S is the length of the path. The induced electric field as a result of a changing magnetic flux is not conservative.

PROBLEM 75 (SCREEN 12.3)

Magnetic Force on a Moving Charge

Problem Description
The magnetic field of the Earth in a certain region is uniform, directed vertically downward, with a magnitude of 0.5×10^{-4} T. A proton is moving horizontally toward the west in this field with a speed of 6.2×10^6 m/s. What are the direction and magnitude of the magnetic force the field exerts on this charge? What is the radius of the circular arc followed by this proton?

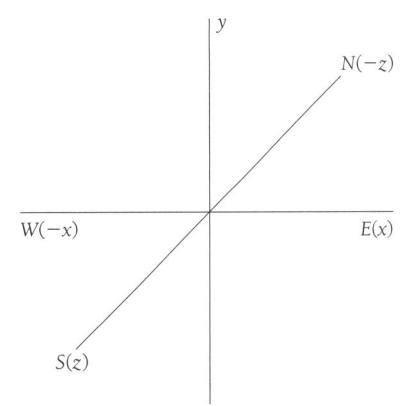

Before we begin...

1. How is magnetic force $\vec{F}_B$ related to the magnetic field $\vec{B}$ and the velocity $\vec{v}$ of a moving charged particle?

2. Does this force change the speed of the particle? If not, what type of acceleration does the particle experience?

3. Draw, not to scale, the magnetic field and velocity on the coordinate axes shown above.

4. Identify the given information:

    magnetic field $B =$

    proton's velocity $v =$

    proton's charge $q =$

    proton's mass $m =$

(continued on next page . . .)

Module 12 **The Magnetic Field**

Solving the problem

5. Substitute the given information into the relationship between $\vec{F}_B$, $\vec{v}$, and $\vec{B}$, and solve for the force $\vec{F}_B$.

6. How is centripetal force related to mass m, speed v, and the radius r of the path?

7. Substitute the given and calculated information into the centripetal force equation to compute the radius of the path.

PROBLEM 76 (SCREEN 12.3)

The Lorentz Force

Problem Description

An electron moves through a uniform electric field $\vec{E}$ with a magnitude of 2.5 V/m in the x direction and 5.0 V/m in the y direction, and a uniform magnetic field $\vec{B}$ = 0.40 T in the z direction. Determine the acceleration of the electron when it has a velocity $\vec{v}$ = 10 m/s in the x direction.

Before we begin...

1. What is the expression for the electric field contribution to the force on the electron?

2. What is the expression for the magnetic field contribution to the force on the electron?

3. How is the acceleration of a particle computed if the net force $\vec{F}$ and the mass m are known?

(continued on next page . . .)

4. Identify the given information:

   x component of the electric field $\vec{E}$ =

   y component of the electric field $\vec{E}$ =

   z component of the electric field $\vec{B}$ =

   electron's velocity $\vec{v}$ =

   electron's charge e =

   electron's mass m_e =

Solving the problem

5. The net force acting upon the charged particle is equal to the vectoral sum of its electrical force $\vec{F}_E$ and its magnetic force $\vec{F}_B$. Evaluate the electric force using $\vec{F}_E = q\vec{E}$.

6. Evaluate the magnetic force $\vec{F}_B$.

7. Add the electric and magnetic force components for each direction to determine the components of the total force.

8. Using the component form of the relation between force, mass, and acceleration, solve for the acceleration in each direction.

Core Concepts in College Physics Workbook

PROBLEM 77 (SCREEN 12.4)

Ampère's Law

Problem Description

A packed bundle of 100 long, straight, insulated wires forms a cylinder of radius $R = 0.50$ cm. Each of the 100 wires carries a current of 2.0 amps "into the page" as shown. What are the magnitude and direction of the magnetic force per unit length on one of the wires that passes through point A on the diagram, at a distance $r = 0.20$ cm from the center of the bundle?

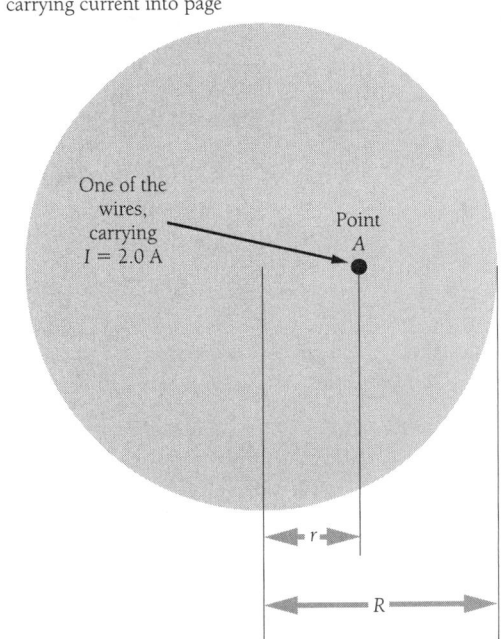

Bundle of 100 wires carrying current into page

One of the wires, carrying $I = 2.0$ A

Point A

Before we begin...

1. What is the expression for the magnetic force per unit length on a long, straight wire carrying a current $\vec{I}$ in a magnetic field $\vec{B}$?

2. Because the bundle is circularly symmetric about its center, Ampère's law should let us find the magnetic field $\vec{B}$ at point A. Draw an amperian loop that passes through point A and takes advantage of the bundle's symmetry.

(continued on next page . . .)

3. What is the total current I_{in} passing through the amperian loop you have drawn?

4. How does Ampère's law relate the magnetic field around your amperian loop to the current I_{in} passing through it?

Solving the problem

5. Use Ampère's law to find the strength of the magnetic field at point A. In which direction does this field point?

6. What are the magnitude and direction of the force per unit length exerted by this field on the wire passing through point A?

PROBLEM 78   (SCREEN 12.5)

Magnetic Flux and Gauss's Law for Magnetism

Problem Description
Consider the cube having a side length *L* as shown in the figure. A uniform magnetic field $\vec{B}$ is directed perpendicular to face *abfe*. Find the magnetic flux through the imaginary planar loops *dfhd* and *acfa*.

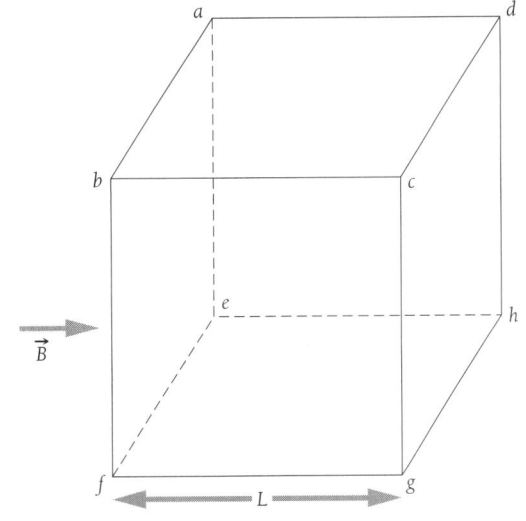

Before we begin...   1. How is magnetic flux calculated?

2. Because $\vec{B}$ is a constant and the surfaces are flat, we need to find the projection of the surfaces onto a surface that has a normal vector parallel to $\vec{B}$—i.e., *abfe* or *dcgh*. The projection of a surface onto another surface is like a surface casting a shadow onto the other. This projection is calculated by $A \cos \theta$, where θ is the angle between the two surfaces. Sketch the imaginary planar loop *dfhd* and its projection onto *dcgh*.

(continued on next page . . .)

Solving the problem

3. Project the surface *dfhd* onto the surface *dcgh*. If the area of *dcgh* is L^2, what is the area of the projection of *dfhd* onto *dcgh*?

   Use this information to compute the magnetic flux through *dfhd*.

4. Repeat the process, projecting *acfa* onto *abfe*.

PROBLEM 79   (SCREEN 12.7)

Faraday's Law of Induction and Lenz's Law

Problem Description

A metal bar spins at a constant rate about one fixed end in the magnetic field of the Earth. The rotation occurs in a region where the component of the Earth's magnetic field perpendicular to the plane of rotation is 3.3×10^{-5} T. If the bar is 1.0 m in length and its angular speed is 5π rad/s, what potential difference is developed between its ends?

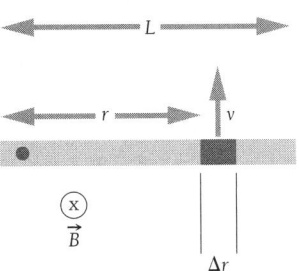

Before we begin...

The *emf* arises from the displacement of free electrons under the influence of the Lorentz force. The electrons move until the electric field created balances the magnetic force.

$$F = qvB \sin\theta + qE = 0$$

Because $\sin\theta = 1$ and $v = \omega r$, where ω is the angular speed and r is the distance along the bar, $\omega r B = -E$. The electric field as a function of r varies linearly from 0 to $B\omega L$, so the total *emf* is

$$\varepsilon = B\omega \frac{L^2}{2}$$

1. Identify the given information:

    magnetic field (component into page) $B =$

    length of bar $L =$

    angular speed of bar $\omega =$

Solving the problem

2. Substitute into the equation for the *emf* of a rotating bar and evaluate the *emf*.

PROBLEM 80 (SCREEN 12.7)

Faraday's Law

Problem Description

A circular metal loop of area 4.0×10^{-4} m² is placed 2 m away from a long, straight wire. At one instant, the wire carries a current of 100 A in the $+y$ direction. At a time 0.083 seconds later, the current in the wire is 100 A in the $-y$ direction. What was the average induced emf in the small loop during this time interval?

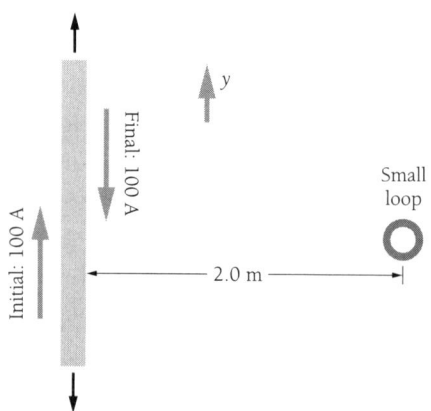

Before we begin...

1. Identify the relevant information:

   area of the loop $A =$

   distance from the wire to the loop $r =$

   initial cuurent in the wire $I_i =$

   final current in the wire $I_f =$

   elapsed time $\Delta t =$

2. Faraday's law states that the induced *emf* in the loop is equal to the rate of change of the magnetic flux through the loop, which may seem rather remote from the information given in this problem. Devise a strategy, identifying the subproblems that you will solve to deduce the *emf* from the given information.

Solving the problem

3. What is the expression for the magnetic field at a distance r from a long, straight wire carrying current I?

4. Use this expression to find the initial and final magnetic fields $\vec{B}_i$ and $\vec{B}_f$ at the position of the small loop.

5. Use your answers to step 4 to find the initial and final magnetic fluxes ϕ_{Bi} and ϕ_{Bf} through the loop. For the purposes of this problem, define flux directed into the page as negative, and flux directed out of the page as positive.

6. What was the net change in magnetic flux $\Delta\phi_B$ during the time interval Δt?

7. Use Faraday's law to find the average induced *emf* in the loop during the time interval Δt. Incidentally, why can we ask only about the average *emf* during that interval, and not about the exact *emf* at any point in time?

MODULE 13   Electric Circuits

INTRODUCTION

In this module, we discuss simple electric circuits. We examine three devices: resistors, capacitors, and inductors. Resistors are nonconservative devices that convert electrical energy into thermal energy. Capacitors and inductors are conservative devices, that store energy in their electric and magnetic fields respectively. The laws of conservation of charge and energy can be employed to analyze circuits containing these devices.

DEFINITIONS

Electric current, I. The rate at which electric charge flows past any point in a conductor. The SI unit for current is the ampere, which is one coulomb per second.

Resistor. A device that resists current flow when a potential difference is applied across the device. The resistance R of an object depends on its size, shape, and temperature, as well as the material of which it is made. The SI unit is the ohm.

Series connection. A connection that occurs when there are no junction points between elements.

Parallel connection. A connection that occurs when the path of a current divides between elements.

Capacitor. A device that stores energy within its electric field. The capacitance C depends upon the geometry and materials of the capacitor. The SI unit of capacitance is the farad, which is one coulomb per volt.

Inductor. A device that stores energy within its magnetic field. The inductance L depends upon the geometry and materials of the inductance coils. The SI unit of inductance is the henry, which is one volt·second per ampere.

Core Concepts in College Physics Workbook

USEFUL EQUATIONS

Ohm's law relates the voltage V to the electric current I and the resistance R by the equation

$$V = IR$$

Resistors connected in series have an equivalent resistance computed by

$$R_S = R_1 + R_2 + \ldots$$

Resistors connected in parallel have an equivalent resistance computed by

$$\frac{1}{R_p} = \frac{1}{R_1} + \frac{1}{R_2} + \ldots$$

Kirchhoff's laws apply the conservation of charge to the junction rule:

$$\sum I_{in} = \sum I_{out}$$

and the conservation of energy to the loop rule:

$$\sum \Delta V = 0 \text{ around the closed loop}$$

Capacitance is related to voltage and charge by the equation

$$Q = CV$$

The energy stored in a capacitor is computed by the relationship

$$U = \frac{1}{2}QV = \frac{1}{2}CV^2 = \frac{Q^2}{2C}$$

Capacitors in series have an equivalent capacitance found by

$$\frac{1}{C_S} = \frac{1}{C_1} + \frac{1}{C_2} + \ldots$$

Capacitors in parallel have an equivalent capacitance computed as

$$C_p = C_1 + C_2 + \ldots$$

Inductance is related to the induced voltage opposing the change in current and the rate of change of electric current by the equation

$$V = L\frac{\Delta I}{\Delta t}$$

The energy stored in an inductor is computed by the relationship

$$U = (1/2)LI^2$$

Inductors in series and parallel have equivalent inductances computed in the same manner as resistors in series and parallel.

An *LC* circuit oscillates with an angular frequency given by

$$\omega_0 = \frac{1}{\sqrt{LC}}$$

An *RLC* circuit oscillates with an angular frequency expressed by

$$\omega = \sqrt{(\omega_0^2 - \beta^2)}$$

where ω_0 is the natural frequency of oscillation for an *LC* circuit and $\beta = R/2L$.

PROBLEM 81 (SCREEN 13.3)

Voltage, Resistance, and Ohm's Law

Problem Description
Batteries are rated in terms of ampere hours (A·h), where a battery rated at 1.0 A·h can produce a current of 1.0 A for 1.0 h. What is the total energy in kilowatt hours, stored in a 12.0 V battery rated at 55.0 A·h? At $0.06 per kilowatt hour, what is the value of the electricity produced by this battery?

Before we begin...

1. From the principles of work and energy, we recall that power is defined as the rate at which work is done or energy is consumed. How is power computed for electric circuits?

   The total energy can be computed by multiplying the power by the time.

2. Identify the given information:

   voltage $V =$

   cost/kW·h $=$

   rating $=$

(continued on next page . . .)

Module 13 **Electric Circuits**

Solving the problem

3. A battery rated at 55.0 A·h can deliver a current of 55.0 A for a time of 1.0 hours. It can also deliver 5.0 A for 11.0 h, or any other product of current and time equaling 55.0 A·h. Write the expression for the energy delivered in terms of current, voltage, and time.

4. What are the dimensions of the equation?

5. By what would we have to multiply the rating of the battery to give the dimensions in the above equation?

6. Use the information given to solve for the energy stored in the battery.

7. Compute the cost of the total energy by multiplying the cost per kW·h by the number of kW·h stored in the battery.

PROBLEM 82 (SCREEN 13.4)

Circuit Analysis and Kirchhoff's Laws

Problem Description
The resistance between terminals a and b in the figure is 75 Ω. If the resistors labeled R have the same value, determine R.

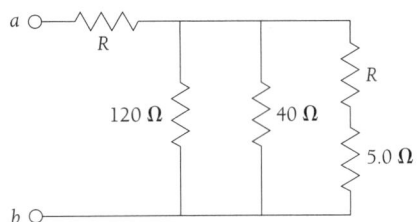

Before we begin...

1. Many times we find that a connection of elements can be simplified by identifying the elements in simple series and in simple parallel. What is the equivalent resistance for resistors connected in simple series?

2. What is the equivalent resistance for resistors connected in simple parallel?

Solving the problem

3. As we attempt to simplify the connection of resistors, observe the two resistors on the far right side of the diagram. How are they connected to each other?

4. Compute and write their equivalent resistance as R_1.

5. Redraw the diagram using R_1.

(continued on next page . . .)

6. In the redrawn diagram, how are R_1, the 120 Ω resistor, and the 40 Ω resistor connected?

   Compute their equivalent resistance as R_2.

7. Redraw the diagram using R_2.

8. In the redrawn diagram, how are R_2 and R connected?

   Write the equation for their equivalent resistance R_E.

9. Because $R_E = 75$ Ω, substitute the value of R_2 and subsequently R_1 into the equation and solve for the value of R.

PROBLEM 83 (SCREEN 13.4)

Circuit Analysis and Kirchhoff's Laws

Problem Description
Find the potential difference between points a and b in the diagram. Find the currents I_1, I_2, and I_3 as shown.

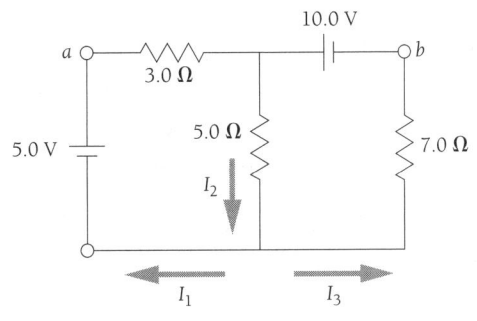

Before we begin...

1. This problem has multiple sources of emf and multiple loops. Solving it requires Kirchhoff's laws. State the junction rule and the loop rule.

Solving the problem

2. Select a junction point in the diagram and apply the junction rule.

There are three unknown currents in this problem, so we need three independent equations to simultaneously solve for them. The junction rule gives us one such equation. The loop rule will give us the other two.

3. Select a loop in the diagram and apply the loop rule. Repeat the process for another loop.

4. Solve the equations for the values of the currents I_1, I_2, and I_3.

5. To evaluate the potential difference between a and b, travel any path between the two and use the algebraic sum of the potentials.

$$V_a - 3(I_1) - 10 = V_b$$
$$-3(I_1) - 10 = \Delta V$$
$$-3(.141) - 10 =$$
$$\Delta V = -10.4 \text{ V}$$

PROBLEM 84   (SCREEN 13.5)

Capacitors

Problem Description

For the system of capacitors shown in the diagram, find the equivalent capacitance of the system, the potential across each capacitor, the charge on each capacitor, and the total energy stored by the group.

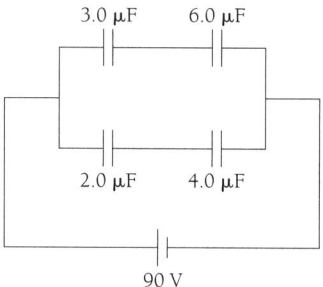

Before we begin...

1. Many times a circuit can be simplified by identifying elements in simple series and in simple parallel. What is the equivalent capacitance for capacitors in series?

2. What is the equivalent capacitance for capacitors in parallel?

3. What do capacitors in series have in common?

   What do capacitors in parallel have in common?

4. What is the relationship between capacitance, charge, and voltage of a capacitor?

5. How is the energy stored in a capacitor computed?

Core Concepts in College Physics Workbook

Solving the problem

6. How are the 3.0 μF capacitor and the 6.0 μF capacitor connected?

   Write the equivalent capacitance of the two as C_2.

7. How are the 2.0 μF capacitor and the 4.0 μF capacitor connected?

   Write the equivalent capacitance of the two as C_3.

8. Redraw the diagram using C_2 and C_3. How are C_2 and C_3 connected?

   Evaluate the equivalent capacitance C_T.

9. Use the answers to the above questions to help evaluate the charge and potential difference across each capacitor.

10. Using the equivalent capacitance, evaluate the energy stored by this capacitor. This is the same as the sum of the energies stored in all of the combined capacitors.

PROBLEM 85 (SCREEN 13.6)

Inductors

Problem Description

Consider the circuit shown in the diagram. What energy is stored in the inductor when the current reaches its final equilibrium value after the switch S is closed?

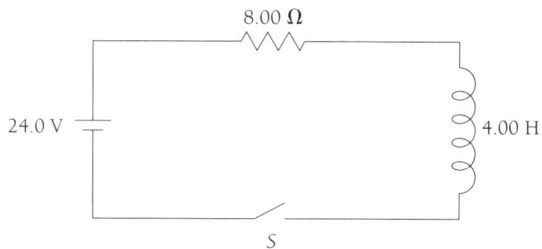

Before we begin...

1. An inductor sets up a potential difference to oppose changes in the current flowing in a circuit. How is this potential difference expressed in terms of the inductance L?

2. Once the circuit has reached equilibrium, what is the rate of change of the current through the inductor?

3. How is the energy stored in the magnetic field of an inductor computed?

4. Identify the given information:

    battery voltage $V =$

    resistance $R =$

    inductance $L =$

Solving the problem

5. Once the opposing potential difference has been reduced to zero, the current flowing through the circuit reaches its maximum value. Use Ohm's law to compute the value of the maximum current.

6. Use the maximum value of the current I and the given inductance L to calculate the energy stored in the magnetic field of the inductor.

Core Concepts in College Physics Workbook

PROBLEM 86   (SCREEN 13.7)

Circuits Containing Resistors, Inductors, and Capacitors

Problem Description
Calculate the inductance of an *LC* circuit that oscillates at a frequency $f = 120$ Hz when the capacitance is 8.00 μF.

Before we begin...   An *LC* oscillating circuit is the electrical equivalent of the mechanical spring-mass system oscillating without friction. From Kirchhoff's loop rule, we obtain

$$L\frac{\Delta I}{\Delta t} + \frac{Q}{C} = 0 \quad \left(\text{with } I = \frac{\Delta Q}{\Delta t}\right)$$

which is analogous to

$$ma + kx = 0 \quad \left(\text{with } a = \frac{\Delta v}{\Delta t},\ v = \frac{\Delta x}{\Delta t}\right)$$

It has already been shown that this mass-spring system will oscillate with an angular frequency

$$\omega_0 = \sqrt{k/m}$$

For the *LC* circuit, direct comparison allows us to predict

$$\omega_0 = \frac{1}{\sqrt{LC}}$$

1. How is the angular frequency ω_0 related to the frequency f?

2. Identify the given information:

   frequency $f =$

   capacitance $C =$

(continued on next page . . .)

Module 13   **Electric Circuits**

Solving the problem

3. Compute the angular frequency ω_0 for this problem.

4. Solve the equation relating the angular frequency to the inductance and capacitance to find the value of the inductance required.

PROBLEM 87   (SCREEN 13.7)

Circuits Containing Resistors, Inductors, and Capacitors

Problem Description

Consider a series LC circuit in which $L = 2.18$ H and $C = 6.00$ nF. What is the maximum value of a resistor that, inserted in series with L and C, allows the circuit to continue to oscillate?

Before we begin...

The description of this problem is that of a damped circuit that has an angular frequency

$$\omega = \sqrt{(\omega_0^2 - \beta^2)}$$

where ω_0 is the natural frequency of oscillation for an LC circuit and $\beta = R/2L$. For small values of R, ω is only slightly less than ω_0.

1. What is the relationship between ω_0, L, and C?

2. Identify the given information:

    inductance $L =$

    capacitance $C =$

Solving the problem

3. Because we wish the circuit to oscillate, β must be less than ω_0. Set the equations for ω_0 and for β equal to each other and solve for the limiting value of R.

MODULE 14

Geometric Optics

INTRODUCTION

Visible light consists of electromagnetic waves with wavelengths so small that noticeable diffraction does not occur as it passes through normal sized openings. In this module, we study the characteristics of light rays that point along the direction of propagation of these waves. We will examine reflection and refraction of light at a boundary between different media and the formation of images using mirrors and thin lenses.

DEFINITIONS

Reflection. When a light ray traveling in a medium reaches a boundary with a second medium, part or all of the ray is reflected back into the first medium. The angle of incidence with respect to the surface normal is equal to the angle of reflection with respect to the surface normal at the point of reflection.

Refraction. When a light ray traveling in a medium reaches a boundary with a second medium, part of the ray may be transmitted into the second medium. As a result of the speed with which light travels through the different mediums, the wavelength of the light will change in the second medium, while the frequency of the light remains unchanged.

Index of refraction. The index of refraction of a material n is the ratio of the speed of light as measured in a vacuum to the speed of light as measured in the material.

Critical angle. The critical angle θ_C is the angle of incidence such that the angle of refraction is exactly 90°.

Lateral magnification. The lateral magnification of an image is the ratio of the height of an image to the height of the object producing the image.

Virtual image. A virtual image is one in which the light rays do not actually converge to the image point, although they appear to diverge from that point.

Focal length. Rays parallel to the principal axis of a mirror or thin lens converge on a single point known as the focal point. The distance from the focal point to the mirror or lens is called the focal length f.

USEFUL EQUATIONS

The law of reflection requires that $\theta_1 = \theta_1'$.

Snell's law for refraction is expressed as

$$n_1 \sin \theta_1 = n_2 \sin \theta_2$$

where θ_2 is the angle with respect to the surface normal that the refracted light ray makes.

The mirror equation and the thin lens equation both are given by

$$\frac{1}{f} = \frac{1}{d_o} + \frac{1}{d_i}$$

with d_o representing the object distance and d_i being the image distance. When d_i is a negative number, the image is virtual and erect; otherwise the image is real and inverted in orientation from the forming object.

The lateral magnification is computed by the equation

$$M = \frac{h_i}{h_o} = -\frac{d_i}{d_o}$$

The lens maker's equation relates the index of refraction of the lens, with respect to the surrounding medium, and the radii of curvature of the lens to the focal length of the lens.

$$\frac{1}{f} = (n-1)\left(\frac{1}{R_1} - \frac{1}{R_2}\right)$$

Module 14   **Geometric Optics**

PROBLEM 88   (SCREEN 14.3)

Reflection

Problem Description
Determine the minimum height of a vertical flat mirror in which a person of height h can see his or her full image.

Before we begin...

1. State the law of reflection.

2. Draw a ray diagram illustrating the problem. Locate the eye position, the top of the head and the feet of the person.

Solving the problem

Use the law of reflection and assume that the vertical distance from the eyes to the top of the head is y_h, while the vertical distance from the eyes to the feet is y_f.

3. How far above the eyes must the mirror top be placed so that the person still can see the top of his head? Draw a ray diagram for this portion of the problem.

4. How far below the eyes must the mirror bottom be placed so that the person can still see his feet? Draw a ray diagram for this portion of the problem.

5. Add these two distances together to get the minimum height of the mirror.

PROBLEM 89 (SCREEN 14.4)

Snell's Law

Problem Description
A light ray in air is incident on a water surface at an angle of 30.0° with respect to the normal to the surface. What is the angle of the refracted ray relative to this normal?

Before we begin...

1. Sketch the problem, labeling all angles.

2. State Snell's law.

3. Identify the given information:

   angle of incidence $\theta_1 =$

   air's index of refraction $n_1 =$

   water's index of refraction $n_2 =$

Solving the problem

4. In this problem, the light is moving from air (n_1) into water (n_2). Apply Snell's law to solve for the angle at which the light will travel with respect to the normal while in the water.

PROBLEM 90 (SCREEN 14.4)

Snell's Law

Problem Description

A glass block having $n = 1.52$ and surrounded by air measures 10.0 cm × 10.0 cm. For an angle of incidence of 45.0°, what is the maximum distance x as shown in the figure so that the ray will emerge from the opposite side?

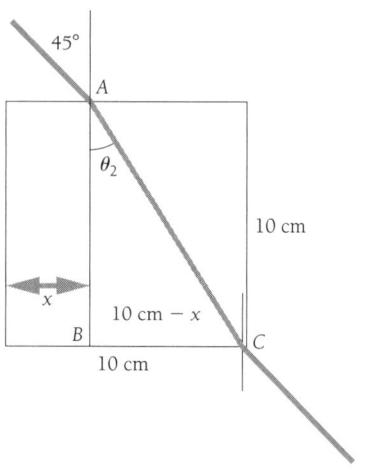

Before we begin...

In this problem we are examining the light during the time that it is traveling through the glass. The angle that the light makes with the normal will be determined by Snell's law and the given information.

1. Identify the given information:

    angle of incidence $\theta_1 =$

    air's index of refraction $n_1 =$

    block's index of refraction $n_2 =$

Solving the problem

2. Use Snell's law and the given information from above to calculate θ_2.

3. Examine triangle ABC in the above diagram. What is the relationship between θ_2 and the sides of the triangle, 10 cm and (10 cm − x)?

4. Use this relationship to solve for x.

Core Concepts in College Physics Workbook

PROBLEM 91 (SCREEN 14.5)

Total Internal Reflection

Problem Description
A fiber optic cable ($n = 1.50$) is submerged in water ($n = 1.33$). What is the critical angle for light to stay inside the cable?

Before we begin...

1. What do we mean by *critical angle*?

2. Identify the given information:

   cable's index of refraction $n_1 =$

   water's index of refraction $n_2 =$

   angle of transmission into water $\theta_2 =$

Solving the problem

3. Apply Snell's law and the definition of the critical angle to compute θ_C.

Module 14 **Geometric Optics**

PROBLEM 92 (SCREEN 14.7)

Flat and Spherical Mirrors

Problem Description

A concave mirror has a focal length of 40.0 cm. Determine the object position for which the resulting image is upright and four times the size of the object.

Before we begin...

1. How is lateral magnification defined? How is it related to the image and object positions?

2. Write the mirror equation.

3. Upright images are of which type, real or virtual? Is the focal length of a concave mirror positive or negative?

4. Identify the given information:

    lateral magnification $M =$

    mirror's focal length $f =$

Solving the problem

5. Use the definition of lateral magnification to give the relationship between the position of the object d_o and the position of the image d_i.

    Substitute into the mirror equation the given information and the relationship between d_o and d_i.

    Evaluate to find d_o.

Core Concepts in College Physics Workbook

PROBLEM 93 (SCREEN 14.8)

Thin Lenses

Problem Description
An object located 32 cm in front of a lens forms an image on a screen 8.0 cm behind the lens. Find the focal length of the lens. Determine the magnification. Is the lens converging or diverging?

Before we begin...

1. If an image is projected onto a screen, is the image real or virtual?

2. State the thin lens equation.

3. How is lateral magnification computed for a thin lens?

4. Identify the given information:

    object distance $d_o =$

    distance to image $|d_i| =$

Solving the problem

5. Use the answers from above to determine whether the image distance d_i is positive or negative.

    Substitute into the thin lens equation to find the focal length f. If the focal length is positive, the lens is converging. If the focal length is negative, the lens is diverging.

6. Substitute the given information into the equation for lateral magnification to solve for M.

Module 14 **Geometric Optics**

SOLUTIONS

Module 2 Vectors

PROBLEM 1 Coordinate Systems

Before we begin...

1.

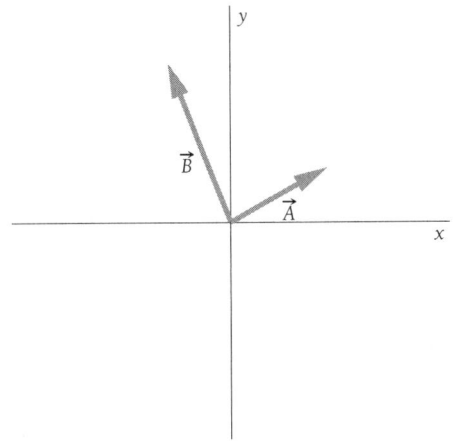

$\vec{A} = (2.50 \text{ m}, 30.0°)$

$\vec{B} = (3.80 \text{ m}, 120.0°)$

2. The cartesian coordinates x and y are related to the polar coordinates, r and θ

$$x = r \cos \theta \quad \text{and} \quad y = r \sin \theta$$

Solving the problem

3. For each vector, $\vec{A}$ and $\vec{B}$, the x and y components are

$$A_x = (2.50 \text{ m}) \cos 30° = 2.17 \text{ m}$$
$$A_y = (2.50 \text{ m}) \sin 30° = 1.25 \text{ m}$$

$$B_x = (3.80 \text{ m}) \cos 120° = -1.90 \text{ m}$$
$$B_y = (3.80 \text{ m}) \sin 120° = 3.29 \text{ m}$$

$$\vec{A} = (2.17 \text{ m}, 1.25 \text{ m})$$
$$\vec{B} = (-1.90 \text{ m}, 3.29 \text{ m})$$

4. The x and y coordinates of $\vec{B} - \vec{A}$ are

$$(\vec{B} - \vec{A})_x = B_x - A_x = -4.07 \text{ m}$$
$$(\vec{B} - \vec{A})_y = B_y - A_y = 2.04 \text{ m}$$

5. The magnitude of $\vec{B} - \vec{A}$ is

$$|\vec{B} - \vec{A}| = \sqrt{4.07^2 + 2.04^2} = 4.55 \text{ m}$$

PROBLEM 2 Vector Addition and Subtraction

Before we begin...

1.
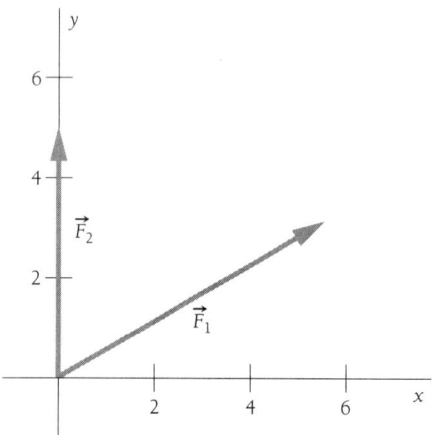

Solving the problem

2.
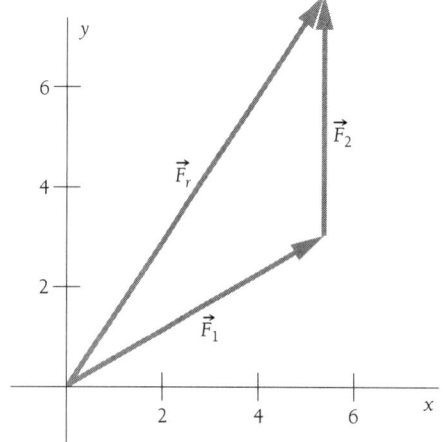

The resultant force is found by translating the tail of $\vec{F}_2$ to the head of $\vec{F}_1$. By measuring it with a ruler and a protractor, we find that the resultant force $\vec{F}_r$ is 9.54 units directed 57° above the x axis.

PROBLEM 3 Vector Components

Before we begin...

1. The rectangular coordinates of a vector are related to the polar coordinates by

$$A_x = A \cos \theta \quad \text{and} \quad A_y = A \sin \theta$$

2.
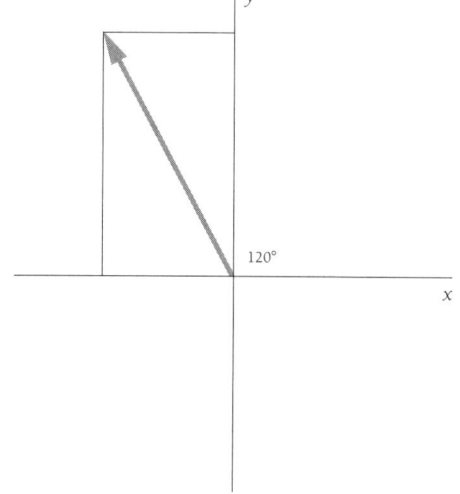

Solving the problem

3. The rectangular coordinates A_x and A_y are

$$A_x = (50.0 \text{ m}) \cos 120° = -25.0 \text{ m}$$

and

$$A_y = (50.0 \text{ m}) \sin 120° = 43.3 \text{ m}$$

PROBLEM 4 Vector Components

Before we begin...

1.

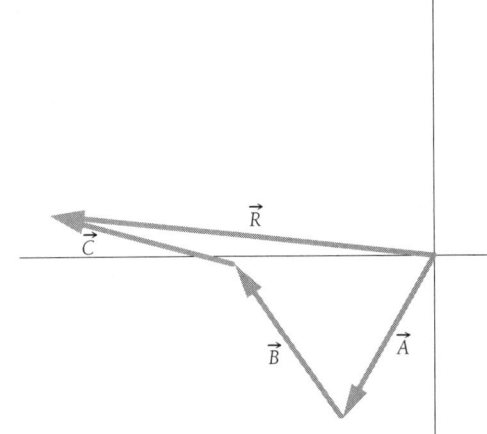

2. The vectors $\vec{A}$, $\vec{B}$, and $\vec{C}$ can be expressed in polar coordinates as $\vec{A} = $ (75 paces, 240°), $\vec{B} = $ (125 paces, 135°), and $\vec{C} = $ (100 paces, 160°).

Solving the problem

3. The x and y components of the three vectors are

$A_x = $ (75 paces) cos 240° $= -37.5$ paces
$A_y = $ (75 paces) sin 240° $= -65.0$ paces

$B_x = $ (125 paces) cos 135° $= -88.4$ paces
$B_y = $ (125 paces) sin 135° $= 88.4$ paces

$C_x = $ (100 paces) cos 160° $= -94.0$ paces
$C_y = $ (100 paces) sin 160° $= 34.2$ paces

4. Adding the x components together and the y components together yields

$R_x = -220$ paces   and $R_y = 57.6$ paces

5. The magnitude of the resultant vector $\vec{R}$ is

$$R = \sqrt{R_x^2 + R_y^2} = 227 \text{ paces}$$

6. The angle that the resultant makes with respect to the x axis is

$$\theta = \tan^{-1}\left(\frac{R_y}{R_x}\right) = -15°$$

Inspection of the signs of the x and y components of R shows that the angle is in the second quadrant, 15° short of the $-x$ axis. Therefore,

$$\theta = 165°$$

PROBLEM 5 The Dot Product

Before we begin...

1.

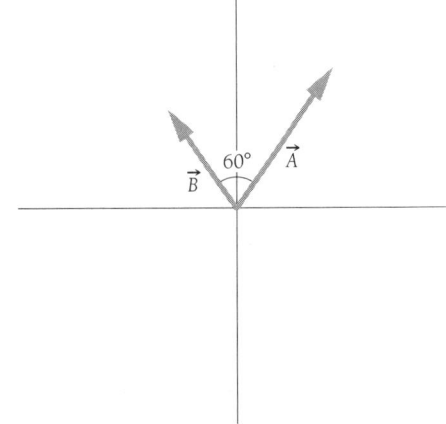

2. The magnitudes of the two vectors are

$|\vec{A}| = 7$ $|\vec{B}| = 4$

3. The resultant of $\vec{A} \cdot \vec{B}$ is a scalar quantity.

Solving the problem

4. $\vec{A} \cdot \vec{B} = 28 \cos 60° = 14$

Module 2 **Vectors**

PROBLEM 6 The Dot Product

Before we begin...

1. This problem expresses the vectors in rectilinear coordinate form, or unit vectors.

2. The x, y, and z components of the vectors are

$$A_x = 0 \qquad A_y = 2.0 \qquad A_z = 0$$
$$B_x = -5.0 \qquad B_y = 3.0 \qquad B_z = 0$$

Solving the problem

3. The scalar product is

$$\vec{A} \cdot \vec{B} = 0 + 6.0 + 0 = 6.0$$

The magnitudes of A and B are

$$A = 2.0 \quad \text{and} \quad B = \sqrt{3.0^2 + 5.0^2} = 5.8$$

4. The angle is

$$\theta = \cos^{-1}\left(\frac{6.0}{2.0 \times 5.8}\right) = 59°$$

Module 3 Kinematics

PROBLEM 7 Displacement, Velocity, and Speed

Before we begin...

1. The acceleration vector is the slope of the velocity as a function of time.

Solving the problem

2. The acceleration is constant from $t = 0$ to $t = 2$ s.

The acceleration in this region is

$$\text{slope} = \frac{(-3 - 0) \text{ m/s}}{(2 - 0) \text{ s}} = -1.5 \frac{\text{m}}{\text{s}^2}$$

3. The acceleration is constant from $t = 2$ to $t = 6$ s.

The value of the acceleration in this region is

$$\text{slope} = \frac{[3 - (-3)] \text{ m/s}}{(6 - 2) \text{ s}} = +1.5 \frac{\text{m}}{\text{s}^2}$$

4. The acceleration is constant from $t = 6$ to $t = 8$ s.

The value of the acceleration in this region is

$$\text{slope} = \frac{(3 - 3) \text{ m/s}}{(8 - 6) \text{ s}} = 0 \frac{\text{m}}{\text{s}^2}$$

5.

(Graph: Acceleration (m/s²) vs Time (s). Horizontal segment at −2 from t=1 to t=3; horizontal segment at +2 from t=4 to t=9.)

6. The average acceleration from $t = 2$ s to $t = 8$ s is

$$\langle \vec{a} \rangle = \frac{[3 - (-3)] \text{ m/s}}{(8 - 2) \text{ s}} = +1.0 \ \frac{\text{m}}{\text{s}^2}$$

7. At $t = 4.0$ s, the acceleration is $+1.5$ m/s².

PROBLEM 8 Displacement, Velocity, and Speed

Before we begin...

1. The given information is

$$\Delta x_1 = +50.0 \text{ m} \qquad \Delta t_1 = 20.0 \text{ s}$$
$$\Delta x_2 = -50.0 \text{ m} \qquad \Delta t_2 = 22.0 \text{ s}$$

The negative value of Δx_2 indicates that she was moving in the $-x$ direction during the second half of her swim.

2. Since both legs of the swim covered the same distance, but the first leg took less time, the swimmer must have been moving faster during the first half.

Solving the problem

3. $\langle v \rangle = \dfrac{\Delta x}{\Delta t}$

4. $\langle v_1 \rangle = \dfrac{\Delta x_1}{\Delta t_1} = +2.50$ m/s

$\langle v_2 \rangle = \dfrac{\Delta x_2}{\Delta t_2} = -2.27$ m/s

As predicted in step 2, $|v_1| > |v_2|$. Again, the opposite signs on these answers reflect the fact that she was moving in opposite directions during the two legs of her trip.

5. $\Delta x_{\text{tot}} = \Delta x_1 + \Delta x_2 = 0.0$ m
(In other words, her trip ended at the same point where it began.)

$$\Delta t_{\text{tot}} = \Delta t_1 + \Delta t_2 = 42.0 \text{ s}$$

6. $\langle v_{\text{tot}} \rangle = \dfrac{\Delta x_{\text{tot}}}{\Delta t_{\text{tot}}} = 0$ m/s

Although the swimmer was moving the entire time, her motion in the $+x$ direction during the first half of her swim "canceled out" her motion in the $-x$ direction during the second half. On the average, she had no net motion in either direction, and ended up back at her starting point.

PROBLEM 9 One-Dimensional Motion at Constant Acceleration

Before we begin...

1. The given information is

Bicyclist A:
$x_{0A} = 0$
$v_{0A} = 0$
$a_A = 2.0 \text{ m/s}^2$

Bicyclist B:
$x_{0B} = 0$
$v_{0B} = 8.0 \text{ m/s}$
$a_B = 0$

2. The only combination of a_A (in m/s²) and v_{0B} (in m/s) that yields units of time (in s) is v_{0B}/a_A. The only combination of a_A (in m/s²) and v_{0B} (in m/s) that yield units of velocity (in m/s) is simply v_{0B} itself. Thus we predict that the time t required for A to catch B, and her velocity v_A at the moment she passes him, will have forms on the order of

$$t \sim \frac{v_{0B}}{a_A} \quad \text{and} \quad v_A \sim v_{0B}$$

(possibly multiplied by a dimensionless constant in each case). Note that we've predicted that A's velocity when she catches B is independent of her acceleration (although the time she takes to catch him will depend on a_A).

Solving the problem

3. When acceleration is equal to a constant a, the position as a function of time is

$$x = x_0 + v_0 t + \frac{1}{2}at^2$$

4. Keeping only the terms which are non-zero for each cyclist,

$$x_A = \frac{1}{2}a_A t^2 \qquad x_B = v_{0B} t$$

5. Setting x_A and x_B equal to each other yields the equation

$$\frac{1}{2}a_A t^2 = v_{0B} t$$

Solving this for t gives two solutions:

$$t = 0 \quad \text{and}$$

$$t = 2\left(\frac{v_{0B}}{a_A}\right) = 2\left(\frac{8 \text{ m/s}}{2 \text{ m/s}}\right) = 8 \text{ s}$$

These two solutions represent the two times when A and B passed each other: B passed A at $t = 0$, and she caught up with him 8 seconds later.

6. The fact that B's velocity v_{0B} appears in the numerator of the equation for t implies that the higher his velocity, the more time it will take A to catch up with him. The fact that her acceleration a_A appears in the denominator implies that the higher her rate of acceleration, the less time it will take her to catch B. Both of these claims make perfect physical sense; if they didn't, then you would know you had made a mistake in deriving the equation.

7. Since A's acceleration is constant and her initial velocity $v_{0A} = 0$, her velocity at any later time is given simply by $v_A = a_A t$.

Substituting in the expression for the time t when she catches B, we find her velocity at that instant:

$$v_A = a_A t = a_A \left(\frac{2v_{0B}}{a_A}\right) = 2v_{0B} = 16 \text{ m/s}$$

8. Except for a dimensionless factor of 2 in each case, the expressions for t and v_A derived in steps 5 and 7 have the same form as the predictions based on dimensional analysis from step 2.

PROBLEM 10 One-Dimensional Motion at Constant Acceleration

Before we begin...

1. To make the units on all given quantities match, you must convert the final velocity from km/s to m/s.

    $v_0 = 0$ m/s

    $v_f = 1.097 \times 10^4$ m/s

    $\Delta x = 220$ m

Solving the problem

2. The best equation to use is the one that relates velocities, distance, and acceleration, without referring directly to time

    $v_f^2 = v_0^2 + 2a(\Delta x)$

3. In this case, $v_0 = 0$, so the equation simplifies to $v_f^2 = 2a\Delta x$. Solving for a,

    $a = \dfrac{v_f^2}{2\Delta x}$

    $= \dfrac{(1.097 \times 10^4 \text{ m/s})^2}{2(220 \text{ m})} = 2.74 \times 10^5$ m/s^2

4. The acceleration experienced by these astronauts is about 27,000 times the familiar free-fall acceleration g. This is, to say the least, an unhealthy situation for humans. Even an unmanned craft would be destroyed by the stress of such a launch, which is probably why you never see this method tried in real life.

5. The most direct link between changing velocity and acceleration is the equation

    $v_f = v_0 + at$

6. Solving for t yields

    $t = \dfrac{v_f - v_0}{a}$

    $= \dfrac{1.097 \times 10^4 \text{ m/s}}{2.74 \times 10^5 \text{ m/s}^2} = 0.040$ s

7. The remaining unused kinematics equation is

    $\Delta x = v_0 t + \dfrac{1}{2} a t^2$

    or ($\Delta x = \frac{1}{2} a t^2$, because $v_0 = 0$. We can check the consistency of our answer by substituting the values of a and t into this equation.

    $\Delta x = \dfrac{1}{2}(2.74 \times 10^5 \text{ m/s})(0.040 \text{ s})^2 = 219$ m

    To within a slight round-off error, this is exactly the length of the cannon given in the problem, so our answer is self-consistent. If we had made any errors in math or logic, the consistency would almost surely have been spoiled, so we would catch our mistake at this point.

PROBLEM 11 Projectile Motion

Before we begin...

1. The given information is

$$x_0 = 0, y_0 = 0$$
(The ball is kicked from the origin.)

$$v_0 = 20.0 \text{ m/s}$$

$$\theta = 53°$$

$$x_G = 36.0 \text{ m} \qquad y_G = 3.05 \text{ m}$$

2. The kinematics equations for freefall are simply the general kinematics equation for constant acceleration, in the special case that $a_x = 0$, $a_y = -g$. With the added simplification that $x_0 = y_0 = 0$, we have

$$x = v_{x0}t \qquad\qquad v_x = v_{x0}$$
$$y = v_{y0}t - \frac{1}{2}gt^2 \qquad v_y = v_{y0} - gt$$

Solving the problem

3. Since the ball is initially moving upward and toward the goal, both the x and y components of its initial velocity are positive. To find their exact values, use trigonometry:

$$v_{x0} = v_0 \cos\theta$$
$$= (20.0 \text{ m/s})(0.601) = 12.0 \text{ m/s}$$

$$v_{y0} = v_0 \sin\theta$$
$$= (20.0 \text{ m/s})(0.798) = 16.0 \text{ m/s}$$

4. Solving the equation $x = v_{x0}t$ for time yields

$$t = \frac{x}{v_{x0}}$$

At the moment the ball passes the goalpost, $x = x_G$, so this must occur at time

$$t = \frac{x_G}{v_{x0}} = \frac{36.0 \text{ m}}{12.0 \text{ m/s}} = 3.0 \text{ s}$$

5. We now know that the ball crosses the goalpost three seconds after being kicked. Its vertical height y at this point in time is

$$y = v_{y0}t - \frac{1}{2}gt^2$$
$$= (16.0 \text{ m/s})(3.0 \text{ s}) - \frac{1}{2}(9.8 \text{ m/s}^2)(3.0 \text{ s})^2$$
$$= 3.9 \text{ m}$$

The ball is 3.9 meters above the ground when it crosses the goalpost. Since the crossbar is only 3.05 meters off the ground, the ball clears the crossbar with almost a meter to spare.

6. The ball's vertical velocity component at any time t is given by $v_y = v_{y0} - gt$. At $t = 3.0$ s, when the ball crosses the goalpost,

$$v_y = v_{y0} - gt$$
$$= (16.0 \text{ m/s}) - (9.8 \text{ m/s}^2)(3.0 \text{ s})$$
$$= -13.4 \text{ m/s}$$

Since v_y is negative, the ball is descending as it clears the crossbar.

PROBLEM 12 Projectile Motion

Before we begin...

1. The given information is

   $\Delta x = 80$ m

   $\Delta y = -35$ m (negative because the vertical motion is downward)

   v_{0x} = unknown

   $v_{0y} = 0$ (the ball was thrown horizontally)

   $g = 9.8$ m/s^2

2.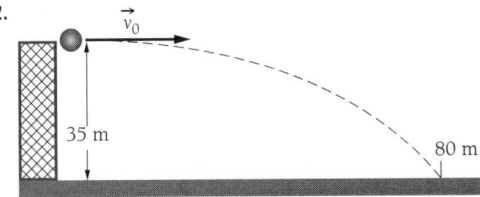

Solving the problem

3. The standard equations of projectile motion are simply the constant-acceleration kinematics equations, in the special case that $a_x = 0$ and $a_y = -g$:

   $v_x = v_{0x}$ $v_y = v_{0y} - gt$

   $\Delta x = v_{0x} t$ $\Delta y = v_{0y} t - \frac{1}{2} g t^2$

   where $\Delta x = x - x_0$, $\Delta y = y - y_0$

4. Because we know the values of Δy, v_{0y}, and g, the equation best suited for finding t is

   $\Delta y = v_{0y} t - \frac{1}{2} g t^2$

   Because $v_{0y} = 0$, this equation reduces to

   $\Delta y = -\frac{1}{2} g t^2$

5. Solving for t,

   $t^2 = -\frac{2\Delta y}{g}$

   $t = \sqrt{-\frac{2\Delta y}{g}}$

   Substituting known values into this expression,

   $t = \sqrt{-\frac{2(-35 \text{ m})}{9.8 \text{ m/s}^2}} = \sqrt{7.14 \text{ s}^2}$

   $= 2.67$ s

6. Now that you know t in addition to the previously known information, you can find v_{0x} using the equation

   $\Delta x = v_{0x} t$

7. Solving for v_{0x},

   $v_{0x} = \frac{\Delta x}{t}$

   Substituting known values into this expression,

   $v_{0x} = \frac{80 \text{ m}}{2.67 \text{ s}}$

   $v_{0x} = 29.9$ m/s

8. The ball's initial velocity has a vertical component $v_{0y} = 0$ and a horizontal component $v_{0x} = 29.9$ m/s. Thus, the vector $\vec{v}_0$ has magnitude 29.9 m/s, directed to the right.

9. The two remaining projectile-motion equations are

   $v_x = v_{0x}$ $v_y = v_{0y} - gt$

   We know that the ball hits the ground at $t = 2.67$ s, so the velocity components at this instant are

   $v_x = 29.9$ m/s

   $v_y = 0 - (9.8 \text{ m/s}^2)(2.67 \text{ s}) = -26.2$ m/s

PROBLEM 13 Uniform Circular Motion

Before we begin...

1. Before the string breaks,

    $r = 0.30$ m

    After the string breaks,

    $\Delta x = 2.0$ m

    $\Delta y = -1.2$ m

    $g = 9.8$ m/s^2

2. The problem involves both uniform circular motion and projectile motion in a constant gravitational field.

3. An object moving at speed v around a circle of radius r experiences a centripetal acceleration directed toward the center of the circle with magnitude

    $$a_c = \frac{v^2}{r}$$

    You already know the circle's radius r, but not the speed v at which the stone moves around the circle.

4. The instant before the string breaks, the stone is in circular motion with speed v. The instant after the string breaks, the stone begins its projectile motion with speed v_{0x}. Since no collision or other sharp impulse takes place, the stone's speed cannot change discontinuously, so the two speeds must be the same, $v = v_{0x}$. Since the circular motion was in the horizontal plane, the stone's initial vertical speed v_{0y} is zero.

Solving the problem

5. In both this problem and Problem 12, an object is launched horizontally (that is, with $v_{0y} = 0$) and moves without air resistance under the influence of a constant gravitational acceleration. The only differences are in the heights from which the projectiles are launched and the distances they travel. The same equations will apply to both problems, although the variables will represent different numerical values.

6. From step 4 of Problem 12, the stone's flight time is given by

    $$t = \sqrt{-\frac{2\Delta y}{g}}$$

    $$= \sqrt{-\frac{2(-1.2 \text{ m})}{9.8 \text{ m/s}^2}} = 0.495 \text{ s}$$

    Then, by step 6 of Problem 12, the stone's initial horizontal speed must be

    $$v_{0x} = \frac{\Delta x}{t}$$

    $$= \frac{2.0 \text{ m}}{0.495 \text{ s}} = 4.04 \text{ m/s}$$

7. The stone's speed v during its circular motion is equal to the initial speed of its projectile motion, or 4.04 m/s. The centripetal acceleration has magnitude

    $$a_c = \frac{v^2}{r} = \frac{(4.04 \text{ m/s})^2}{0.30 \text{ m}} = 54.4 \text{ m/s}^2$$

    The direction of the centripetal acceleration vector, of course, is toward the center of the circle at all times.

PROBLEM 14 Relative Motion

Before we begin...

1.

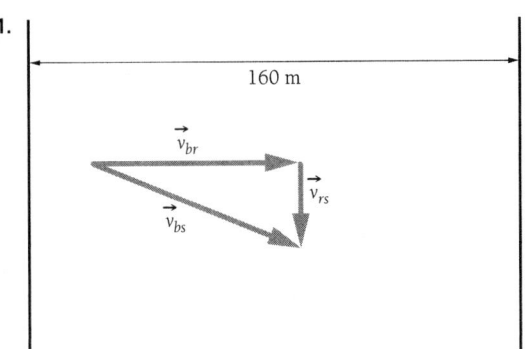

2. The vector sum is important because the current is pushing the boat downstream at the same time the boat is crossing the river.

Solving the problem

3. The magnitude of the velocity of the boat relative to the shore is

$$v_{bs} = \sqrt{v_{br}^2 + v_{rs}^2} = 2.5 \text{ m/s}$$

4. The time required to cross the river is

$$t = \frac{\Delta x}{\langle v_x \rangle} = \frac{160 \text{ m}}{2.00 \text{ m/s}} = 80.0 \text{ s}$$

5. The distance downstream that the boat will travel is

$$\Delta y = \langle v_y \rangle t = (1.5 \text{ m/s})(80 \text{ s}) = 120 \text{ m}$$

Module 4 Forces

PROBLEM 15 Motion, Newton's First Law, and Force

Before we begin...

1. The net force being exerted on the boat must be zero if it moves at constant velocity.

2.

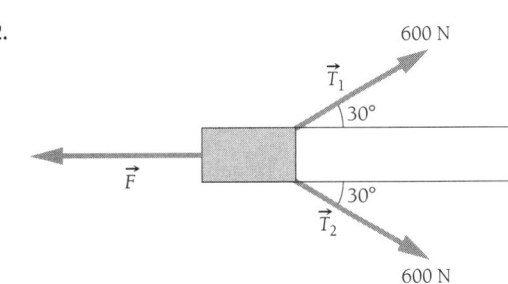

Solving the problem

3. Expressing Newton's first law in component form, with each component sum equal to zero, for x,

$$T_{1x} + T_{2x} + F_x = 0$$

where $T_{1x} = T_1 \cos 30°$ and $T_{2x} = T_2 \sin 30°$.

Substituting for T_1 and T_2,

$$520 \text{ N} + 520 \text{ N} + F_x = 0$$

For y,

$$T_{1y} + T_{2y} + F_y = 0$$

where $T_{1y} = T_1 \sin 30°$ and

$$T_{2y} = -T_2 \sin 30°.$$

Substituting for T_1 and T_2,

$$300 \text{ N} - 300 \text{ N} + F_y = 0$$

4. The x and y components of the resistive force $\vec{F}$ are

$$F_x = -1040 \text{ N} \quad \text{and} \quad F_y = 0$$

Therefore $\vec{F} = 1040$ N in the $-x$ direction.

PROBLEM 16 Newton's Second Law

Before we begin...

1. The given information for this problem is

   $\vec{v}_i = 3.0$ m/s $m_1 = 85$-kg $\Delta t = 0.5$ s

   $\vec{v}_f = 4.0$ m/s $m_2 = 58$-kg

Solving the problem

2. The acceleration that the sprinter must experience is

$$\vec{a} = \frac{\vec{v}_f - \vec{v}_i}{t} = 2.0 \text{ m/s}^2$$

3. Newton's second law allows you to calculate the net force from the mass and the acceleration.

4. The force on the 85-kg sprinter is

$$\vec{F}_{net} = M\vec{a} = (85 \text{ kg})(2.0 \text{ m/s}^2) = 170 \text{ N}$$

5. This same force would cause the 58-kg sprinter to accelerate

$$\vec{a} = (170 \text{ N})(58 \text{ kg}) = 2.93 \text{ m/s}^2$$

PROBLEM 17 Newton's Third Law

Before we begin...

1.

2. Both blocks must have the same acceleration.

Solving the problem

3. Applying Newton's second law to relate the net force on each object to the object's mass and acceleration,

$$F - F_{contact} = m_1 a \quad \text{(block 1)}$$

$$F_{contact} = m_2 a \quad \text{(block 2)}$$

4. Adding the two equations yields

$$F = (m_1 + m_2)a$$

$$a = \frac{F}{m_1 + m_2}$$

5. Substituting our expression for a in the equation for block 2 gives an expression for $F_{contact}$, when the force is applied on block 1:

$$F_{contact} = \frac{Fm_2}{(m_1 + m_2)}$$

6. Since the result should not change with a different viewpoint, it is apparent that situation (b), with the force applied to block 2, looks just like situation (a) viewed from "behind", with m_1 switched with m_2. Nowhere in our analysis have we yet used knowledge about the relative sizes of m_1 and m_2.

Therefore, when the force is applied on block 2,

$$F_{contact} = \frac{Fm_1}{(m_1 + m_2)}$$

Core Concepts in College Physics Workbook

7. With the knowledge that $m_1 > m_2$, we can say that $F_{contact}$ in situation (a) is less than $F_{contact}$ in situation (b).

Following up

Instead of doing the entire analysis again for situation (b), the property of symmetry with respect to reflection was used. Since the relative sizes of m_1 and m_2 did not matter in the derivation of the results, we could then immediately write down the contact force for the second situation.

PROBLEM 18 Newton's Third Law

Before we begin...

1.

2. All three blocks must have the same acceleration. The applied force is pushing on the first block.

Solving the problem

3. The net force on each block is related to the block's mass and acceleration as follows:

Block 1 $F - P_1 = M_1 a$

Block 2 $P_1 - P_2 = M_2 a$

Block 3 $P_2 = M_3 a$

4. Solving the three simultaneous equations for the acceleration yields

$$a = 2.00 \text{ m/s}^2$$

5. The contact force acting upon the object is

Block 3 $P_2 = (4.00 \text{ kg})(2.00 \text{ m/s}^2) = 8.00 \text{ N}$

Block 1 $(18.0 \text{ N}) - P_1 = (2.00 \text{ kg})(2.00 \text{ m/s}^2)$
$= 4.00 \text{ N}$

$$P_1 = 14.0 \text{ N}$$

As a check, the net force on Block 2 must be 6.00 N. Observe that $P_1 - P_2 = 6.00$ N.

Module 4 **Forces**

PROBLEM 19 Free-Body Diagrams

Before we begin...

1. The accelerations must be the same because the string is assumed not to stretch.

2.

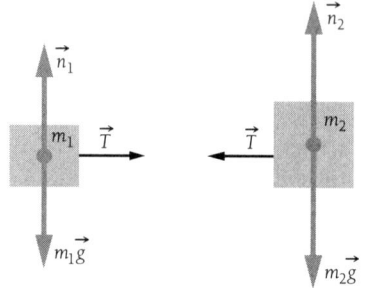

Solving the problem

3. Newton's second law applied to mass 1 is

$$T = m_1 a$$

and applied to mass 2 is

$$F - T = m_2 a$$

4. Adding the two equations together gives us Newton's second law for the system

$$F = (m_1 + m_2)a$$

Solving for the acceleration,

$$a = \frac{F}{m_1 + m_2}$$

$$= \frac{50 \text{ N}}{(10 + 20) \text{ kg}} = 16.7 \text{ m/s}^2$$

5. Substituting our expression for a into our first equation gives an expression for the tension

$$T = m_1 \frac{F}{m_1 + m_2}$$

$$= (50 \text{ N}) \frac{10 \text{ kg}}{(10 + 20) \text{ kg}} = 16.7 \text{ N}$$

6. Adding the kinetic friction force to the free-body diagrams from above,

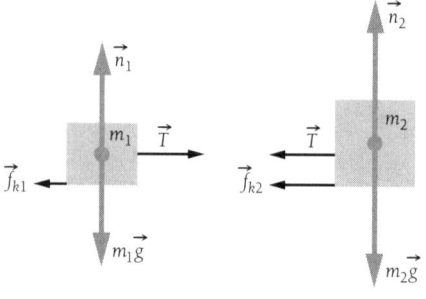

7. With the friction force considered, the accelerations must still be the same.

8. With the kinetic friction force, Newton's second law for mass 1 becomes

$$T - \mu_k m_1 g = m_1 a$$

and that of mass 2 is

$$F - T - \mu_k m_2 g = m_2 a$$

9. Proceeding as before, the two equations added again give us Newton's second law for the system

$$F - \mu_k m_1 g - \mu_k m_2 g = m_1 a + m_2 a$$

Solving for the acceleration,

$$a = \frac{F}{m_1 + m_2} - \mu_k g$$

The first term is the acceleration from above without friction, and the second term is the decrease in the acceleration caused by the friction force.

$$a = 1.67 \text{ m/s}^2 - (0.10)(9.8 \text{ m/s}^2)$$

$$= 0.69 \text{ m/s}^2$$

Core Concepts in College Physics Workbook

10. Solving equation 6 for the tension and substituting our expression for a gives the expression for the tension

$$T = m_1\left(\frac{F}{m_1 + m_2} - \mu_k g\right) + \mu_k m_1 g$$

Note that the third term, the friction force, exactly cancels the second term, because the friction causes the acceleration to be less in a mass-independent way. Thus the expression for tension has exactly the same form as in the frictionless case

$$T = m_1 \frac{F}{m_1 + m_2}$$

$$= (50 \text{ N})\left(\frac{10 \text{ kg}}{(10 + 20) \text{ kg}}\right) = 16.7 \text{ N}$$

PROBLEM 20 Free-Body Diagrams

Before we begin...

1.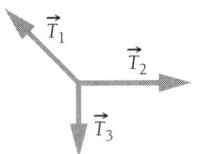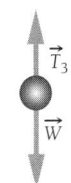

2. The system is in equilibrium; the net force must be zero.

3. The weight of the ball is

$$W = mg = 98.0 \text{ N}$$

Solving the problem

4. The magnitude of the tension T_3 is

$$T_3 = W = 98.0 \text{ N}$$

5. The first free-body diagram shows that the tension $\vec{T}_3$ is pulling vertically downward ($-y$ direction). The directions of $\vec{T}_1$ and $\vec{T}_2$ are

$$T_2 \text{ is at } 0° \text{ and } T_1 \text{ is at } 120°$$

6. Resolving the tensions into their x and y components yields

$$T_{1x} = T_1 \cos 120° = -0.5\, T_1$$
$$T_{1y} = T_1 \sin 120° = +0.866\, T_1$$

$$T_{2x} = T_2 \cos 0° = T_2$$
$$T_{2y} = T_2 \sin 0° = 0$$

$$T_{3x} = T_3 \cos 270° = 0$$
$$T_{3y} = T_3 \sin 270° = -98.0 \text{ N}$$

The condition $T_{1y} + T_{2y} + T_{3y} = 0$ requires

$$0.866 T_1 - 98.0 \text{ N} = 0$$
$$T_1 = 113 \text{ N}$$

The other equilibrium condition, $T_{1x} + T_{2x} + T_{3x} = 0$, yields

$$-0.5 T_1 + T_2 = 0$$
$$T_2 = 0.5 T_1 = 56.5 \text{ N}$$

PROBLEM 21 Free-Body Diagrams

Before we begin...

1. The 4.0-kg object weighs more and so will exert more downward force than the 2.0-kg object. By common sense, the 4.0-kg object will accelerate downward.

2. All of the accelerations must have the same magnitude because the strings are assumed not to stretch.

3. 4.0 kg object   1.0 kg object   2.0 kg object

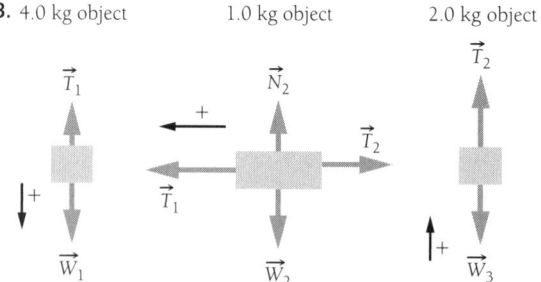

Solving the problem

4. For the three objects, we find

     4.0 kg $W_1 - T_1 = M_1 a$

     1.0 kg $T_1 - T_2 = M_2 a$

     2.0 kg $T_2 - W_3 = M_3 a$

5. Adding all three equations together and solving for a,

$$a = \frac{W_1 - W_3}{M_1 + M_2 + M_3} = \frac{39.2 \text{ N} - 19.6 \text{ N}}{7 \text{ kg}} = 2.8 \text{ m/s}^2$$

6. Substituting this result back into the first and third equations yields

$$T_1 = W_1 - M_1 a$$
$$= 39.2 \text{ N} - (4 \text{ kg})(2.8 \text{ m/s}^2) = 28.0 \text{ N}$$

$$T_2 = W_2 + M_2 a$$
$$= 19.6 \text{ N} + (2 \text{ kg})(2.8 \text{ m/s}^2) = 25.2 \text{ N}$$

As a consistency check, note that this gives a net force on the 1-kg mass of $T_1 - T_2 = 2.8$ N. This is just what is needed to give it the predicted acceleration of 2.8 m/s², so our result is consistent.

PROBLEM 22 Free-Body Diagrams

Before we begin...

1.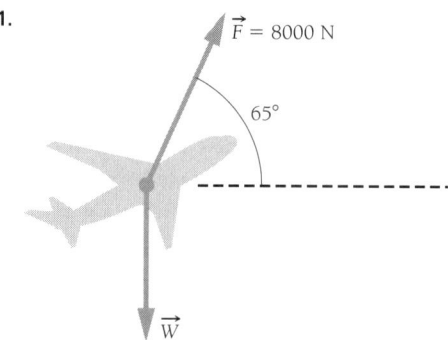

2. Choose the y axis along the vertical, positive direction up, and the x axis along the horizontal, positive in the direction of motion.

Solving the problem

3. The weight has component $-W$ in the y axis. The lift and thrust force $\vec{F}$, of magnitude 8000 N, has components

$$F_x = F \cos 65° \text{ and } F_y = F \sin 65°$$

4. Newton's second law in component form gives the two equations

$$F_x = m a_x$$
$$F_y - W = 0$$

where m is the mass of the plane.

Solving for W and substituting from the components, gives an expression for the weight of the plane

$$W = F_y = F \sin 65° = (8000 \text{ N}) \sin 65°$$

5. Solving the equation $W = mg$ for m and substituting the result and our other components, yields

$$F_x = F \cos 65° = W \frac{a_x}{g}$$

or

$$a_x = \frac{gF \cos 65°}{W}$$

$$= \frac{gF \cos 65°}{F \sin 65°} = \frac{g \cos 65°}{\sin 65°} = \frac{g}{\tan 65°}$$

Following up

Is it a surprise that the horizontal acceleration of the plane is independent of the total force? If a second plane had twice the mass and twice the force in the same direction, it too would have the same vertical and horizontal accelerations.

PROBLEM 23 Centripetal Force

Before we begin...

1.

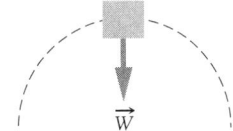

2. If the driver is on the verge of losing contact with the seat, then the seat is exerting no normal force on her. A normal force does not exist between the driver and the seat because the weight of the driver is providing exactly the centripetal force.

Solving the problem

3. The net force providing the centripetal force is

$$W = mg = F_c = \frac{mv^2}{r}$$

4. The speed of the car is

$$v = \sqrt{rg} = 13.3 \text{ m/s}$$

5. The mass appears in both the net force and the centripetal force. Both sides of the equation can be divided by the mass without altering the relationship.

PROBLEM 24 Fictitious Forces: Motion in Accelerated Reference Frames

Before we begin...

1.

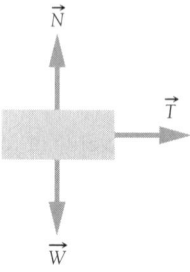

Solving the problem

2. The acceleration of the mass with respect to the inertial frame of reference is

$$\vec{T} = m\vec{a}, \quad \text{so}$$

$$\vec{a} = \frac{\vec{T}}{m} = \frac{18.0 \text{ N}}{5.00 \text{ kg}}$$

$$= 3.60 \text{ m/s}^2 \text{ in the } +x \text{ direction}$$

3. When $\vec{a} = 0$, $\vec{T}$ also must be zero.

4. In the noninertial frame of reference, the net force appears to be zero. For this to be accomplished, we must introduce a fictitious force acting to the left and equal in magnitude to the tension $\vec{T}$.

Module 5 Work and Energy

PROBLEM 25 Work

Before we begin...

1. The relevant information is

$$m = 5.0 \text{ kg} \qquad \theta = 30°$$
$$s = \Delta x = 2.5 \text{ m} \qquad t = 2.0 \text{ s}$$
$$v_0 = 0$$

2.

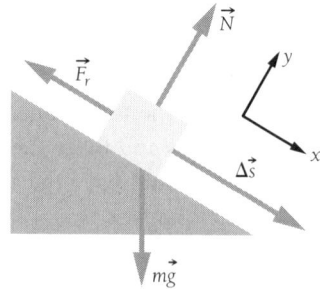

Solving the problem

3. The angles are

angle between $\Delta \vec{s}$ and $m\vec{g} = 60°$

(Because $\Delta \vec{s}$ makes a 30° angle with the horizontal, it makes a 60° angle with a vertical vector such as $m\vec{g}$.)

angle between $\Delta \vec{s}$ and $\vec{N} = 90°$

(The displacement $\Delta \vec{s}$ points along the ramp's surface, while normal force is always perpendicular to the surface.)

angle between $\Delta \vec{s}$ and $\vec{f} = 180°$

(The frictional force $\vec{f}$ points directly backward along the block's path.)

4. The general equation for work done on an object by any force is $W = \vec{F} \cdot \Delta\vec{s}$.

Work done on block by gravity:

$$W = \vec{F} \cdot \Delta\vec{s} = (mg)(\Delta s) \cos 60°$$

$$W = (49 \text{ N})(2.5 \text{ m})(0.500) = 61.3 \text{ J}$$

Work done on block by normal force N:

$$W = \vec{F} \cdot \Delta\vec{s} = (N)(\Delta s) \cos 90° = 0$$

Because $\cos 90° = 0$, we do not need to know the magnitude of the normal force to know that it does no work on the block. This is a useful general rule: forces which act perpendicular to an object's direction of motion never do any work on that object.

5. The most useful kinematics equation for this situation is

$$\Delta x = v_0 t + \frac{1}{2}at^2$$

In this case, the block's initial velocity $v_0 = 0$, so the equation reduces to

$$\Delta x = \frac{1}{2}at^2$$

Solving for a,

$$a = \frac{2\Delta x}{t^2}$$
$$= \frac{2(2.5 \text{ m})}{(2.0 \text{ s})^2} = 1.25 \text{ m/s}^2$$

6. Three of the four vectors point directly along a coordinate axis.

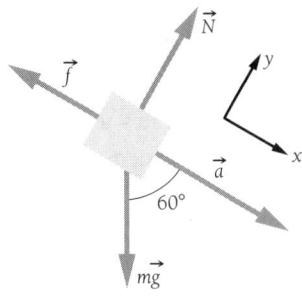

$(mg)_x = mg \cos 60°$
$(mg)_y = -mg \sin 60°$

$N_x = 0 \quad N_y = N$

$f_x = -f \quad f_y = 0$

$a_x = a \quad a_y = 0$

7. Both the block's motion and the unknown force $\vec{f}$ lie entirely along the x axis in our chosen coordinate system, so we will concentrate on the x components when applying Newton's second law:

$$F_{\text{net-}x} = ma_x$$

$$N_x + (mg)_x + f_x = ma_x$$

Substituting the vector components from step 6,

$$mg \cos 60° - f = ma$$

Solving for the friction $\vec{f}$,

$$f = mg \cos 60° - ma = m(g \cos 60° - a)$$
$$= (5.0 \text{ kg})[(9.8 \text{ m/s}^2)(0.500) - (1.25 \text{ m/s}^2)]$$
$$= 18.25 \text{ N}$$

(continued on next page . . .)

8. Recalling that the frictional force makes a 180° angle with the block's displacement, the work it does on the block is given by

$$W = \vec{F} \cdot \vec{\Delta s} = f(\Delta s) \cos 180°$$
$$= (18.25 \text{ N})(2.50 \text{ m})(-1.0) = -45.6 \text{ J}$$

Again, this result illustrates a useful general rule: when the force is directly opposed to the object's direction of motion, the work done is negative force times distance.

9. The block slid down a 2.5-meter incline in 2.0 seconds, so its average speed was

$$\langle v \rangle = \frac{\Delta x}{\Delta t} = \frac{2.5 \text{ m}}{2.0 \text{ s}} = 1.25 \text{ m/s}$$

For any uniformly accelerated object, it's also true that

$$\langle v \rangle = \frac{1}{2}(v_0 + v_f)$$

In this case, the initial velocity $v_0 = 0$, so we can easily find the block's final velocity v_f:

$$\langle v \rangle = \frac{1}{2} v_f$$

$$v_f = 2\langle v \rangle = 2(1.25 \text{ m/s}) = 2.50 \text{ m/s}$$

Thus, the block began with no kinetic energy and ended up with kinetic energy

$$K_f = \frac{1}{2} m v_f^2$$

$$= \frac{1}{2}(5.00 \text{ kg})(2.50 \text{ m/s})^2 = 15.6 \text{ J}$$

According to the work-energy theorem, the net work done by all forces acting on the block gives the change in the block's kinetic energy

$$W_{net} = +15.6 \text{ J}$$

We already know that gravity performed 61.3 J of work on the block, and the normal force performed none, so the work done by friction must have been

$$15.6 \text{ J} - 61.3 \text{ J} = -45.7 \text{ J}$$

Except for a slight roundoff difference, this is exactly the answer we found in step 9, but, by using the work-energy theorem, we've avoided any need to directly calculate the frictional force.

Core Concepts in College Physics Workbook

PROBLEM 26 Important Examples of Work: Gravity and Springs

Before we begin...

1. The cheerleader has to exert a force equal to the weight of his partner.

$$F = mg = (50.0 \text{ kg})(9.80 \text{ m/s}^2)$$
$$= 490 \text{ N}$$

2. The displacement is $\Delta y = 0.60$ m.

Solving the problem

3. The work done for each replication is

$$W = \vec{F} \cdot \vec{s} = Fs \cos \theta$$

Because the force and the displacement are in the same direction, $\cos \theta = 1.00$. For each time the partner was lifted,

$$W = (490 \text{ N})(0.60 \text{ m}) = 294 \text{ J}$$

4. The total work done is

$$20 \times 294 \text{ J} = 5.88 \times 10^3 \text{ J}$$

PROBLEM 27 Energy

Before we begin...

1. The relevant information is

$$m_{car} = 1000 \text{ kg} \qquad m_{truck} = 4000 \text{ kg}$$
$$v_{truck} = 60 \text{ mph}$$

It is not necessary to convert the truck's speed into m/s, because we will eventually find the ratio of the car's speed to that of the truck. The ratio of two quantities with the same units is dimensionless and doesn't depend on the units they're measured in. Still, unless you're certain that the only way a variable will be used in a problem is in a dimensionless ratio, the safest thing to do is convert it into standard units. In this case, you would find $v_{truck} = 26.9$ m/s.

Solving the problem

2. The formula for kinetic energy is

$$K = \frac{1}{2}mv^2$$

3. In order for the car and the truck to have the same kinetic energy,

$$\frac{1}{2}m_{car}v_{car}^2 = \frac{1}{2}m_{truck}v_{truck}^2$$

4. The factors of $1/2$ cancel immediately; then,

$$\frac{v_{car}^2}{v_{truck}^2} = \frac{m_{truck}}{m_{car}}$$

$$\frac{v_{car}}{v_{truck}} = \sqrt{\frac{m_{truck}}{m_{car}}} = \sqrt{4.0} = 2.0$$

In other words,

$$v_{car} = 2.0 v_{truck} = 120 \text{ mph}$$

Module 5 **Work and Energy**

5. If the driver had been going the same speed as the truck, then his car would have had one-fourth the truck's kinetic energy, since kinetic energy is directly proportional to mass. But he claims he had the same kinetic energy as the truck—in other words, four times the energy he would have had driving at 60 mph. Because an object's kinetic energy is proportional to the square of its speed, doubling its speed will give it four times the kinetic energy. So if the driver is going 120 mph, he will have four times the kinetic energy he would have at 60 mph; as we've just argued, this is exactly the same as the truck's kinetic energy.

PROBLEM 28 Conservative Forces

Before we begin...

1. The relevant information is

$m = 4.0$ kg $\qquad g = 9.8$ m/s^2
$h_{shelf} = 2.0$ m $\qquad h_{floor} = 6.0$ m

Solving the problem

2. Relative to the floor,

$$h_1 = 0$$
$$h_2 = h_{shelf} = 2.0 \text{ m}$$

In other words, before jumping, the cat is on the floor, and after jumping she is two meters above it.

3. In a constant gravitational field of strength g, an object's gravitational potential relative to some reference point is

$$U = mgh$$

where m is the object's mass and h is its height above the reference point (or below it, if h is negative). In this case, the reference point is the floor of the apartment.

4. Relative to the floor,

Before jumping:
$U_1 = mgh_1$
$\quad = (4.0 \text{ kg})(9.8 \text{ m/s}^2)(0) = 0$

After jumping:
$U_2 = mgh_2$
$\quad = (4.0 \text{ kg})(9.8 \text{ m/s}^2)(2.0 \text{ m}) = 78$ J

5. Relative to the street,

$$h_1 = h_{floor} = 6.0 \text{ m}$$
$$h_2 = h_{floor} + h_{shelf} = 8.0 \text{ m}$$

In other words, before jumping, the cat is on the floor, 6.0 meters above the street. After jumping onto the shelf, she is two meters higher, or 8.0 meters above the street.

Core Concepts in College Physics Workbook

6. Relative to the street,

   Before jumping:
   $$U_1 = mgh_1$$
   $$= (4.0 \text{ kg})(9.8 \text{ m/s}^2)(6.0 \text{ m}) = 235 \text{ J}$$

   After jumping:
   $$U_2 = mgh_2$$
   $$= (4.0 \text{ kg})(9.8 \text{ m/s}^2)(8.0 \text{ m}) = 313 \text{ J}$$

7. The cat's potential energy relative to the floor measures how much work must be done to move her from the floor to her present location. Her energy relative to the street measures the work that would have to be done to move her from street level to her present location. Thus, for example, her potential energy U_2 while sitting on the bookshelf is much higher when measured relative to the street than when measured relative to the floor; it takes more work to move a cat from the street to a shelf in a third-floor apartment than it does to move her from the floor of the apartment to the shelf.

However, whether measured with respect to the floor or to the shelf, the difference between U_1 when on the floor and U_2 when on the shelf is always the same:

   Relative to floor,
   $$\Delta U = U_2 - U_1 = 78 \text{ J} - 0 \text{ J} = 78 \text{ J}$$

   Relative to street,
   $$\Delta U = U_2 - U_1 = 313 \text{ J} - 235 \text{ J} = 78 \text{ J}$$

It is the change of energy from one state to another that is physically interesting, not the total amount of energy present. So the choice of reference points for potential energy is purely a matter of convenience; no matter which reference point you choose, the change in energy remains the same.

PROBLEM 29 The Work-Energy Theorem

Before we begin...

1. The relevant information is

   $m = 0.020$ kg $y_1 = -0.120$ m
   $y_2 = 0.0$ m $y_3 = 20.0$ m

   The values y_1, y_2, and y_3 represent the projectile's displacement from the spring's equilibrium position in each of the three states: 12 centimeters below equilibrium, exactly at equilibrium, and 20 meters above equilibrium, respectively.

Solving the problem

2. In this state, the projectile is not yet moving, so it has no kinetic energy. As it is below the chosen zero point for gravitational potential by a distance y_1, it does have gravitational potential energy mgy_1. Also, the spring is compressed by a distance y_1 from its equilibrium position, and thus stores potential energy $\frac{1}{2}ky_1^2$. Thus, the system's total energy in State #1 is

$$E_1 = mgy_1 + \frac{1}{2}ky_1^2$$

3. As the projectile reaches the top point in its flight, it is momentarily at rest, so it has no kinetic energy. Nor is any energy stored in the spring. The only energy present in this state is the gravitational potential energy of the projectile, which is a distance y_3 above the zero point of gravitational potential:

$$E_3 = mgy_3$$

4. Assuming that friction and air resistance are negligible, the only forces acting on the projectile are gravity and the spring force, both of which are conservative (that is, the work they do can be incorporated into potential energy). When no nonconservative forces perform work on a system, the system's total mechanical energy remains constant, so E_1 and E_3 should be equal.

(continued on next page . . .)

5. Equating the expressions for E_1 and E_3,

$$mgy_1 + \frac{1}{2}ky_1^2 = mgy_3$$

6. The only unknown quantity in the above equation is the spring constant k. Solving for k,

$$\frac{1}{2}ky_1^2 = mgy_3 - mgy_1 = mg(y_3 - y_1)$$

$$k = \frac{2mg(y_3 - y_1)}{y_1^2}$$

$$= \frac{2(0.020 \text{ kg})(9.8 \text{ m/s}^2)(20.12 \text{ m})}{(0.12 \text{ m})^2}$$

$$= 548 \text{ N/m}$$

7. As the projectile passes equilibrium, the spring is no longer compressed, so it stores no potential energy. The projectile's gravitational potential energy is also zero, because it is at our chosen reference point for gravitational potential. The projectile is moving with speed v, so it does possess kinetic energy:

$$E_2 = \frac{1}{2}mv^2$$

8. For the same reasons argued in step 4, the system's total mechanical energies in State #2 and State #3 should be equal. Thus, equating the expressions for E_2 and E_3,

$$\frac{1}{2}mv^2 = mgy_3$$

Solving this equation for v, we find the projectile's speed as it passes the equilibrium point:

$$v = \sqrt{2gy_3}$$
$$= \sqrt{2(9.8 \text{ m/s}^2)(20.0 \text{ m})} = \sqrt{392 \text{ m}^2/\text{s}^2}$$
$$= 19.8 \text{ m/s}$$

PROBLEM 30 The Work-Energy Theorem

Before we begin...

1. The relevant information is

$m = 80.0$ kg
$y_1 = 6.0$ m
$v_1 = 0$
$y_2 = 0.0$ m
$v_2 = 5.0$ m/s

The values y_1 and y_2 represent the firefighter's displacement from the bottom of the pole in each of the two states: 6.0 meters above the base of the pole, and exactly at the base of the pole, respectively.

Solving the problem

2. Since the firefighter is higher than our chosen reference point at the bottom of the pole, he has gravitational potential energy mgy_1. He is not yet moving, so he has no kinetic energy. Thus, the total energy in state #1 is

$$E_1 = mgy_1$$

3. As the firefighter reaches the bottom of the pole, he is moving at a nonzero speed v_2, so he has kinetic energy. His gravitational potential energy is zero, because the base of the pole is our chosen reference point for gravitational potential. Thus,

$$E_2 = \frac{1}{2}mv_2^2$$

4. Substituting known values into the equations for E_1 and E_2, we find

$$E_1 = (80.0 \text{ kg})(9.8 \text{ m/s}^2)(6.0 \text{ m}) = 4700 \text{ J}$$
$$E_2 = \frac{1}{2}(80.0 \text{ kg})(5.0 \text{ m/s}^2) = 1000 \text{ J}$$

The change in mechanical energy is

$$\Delta E = E_2 - E_1 = -3700 \text{ J}$$

That is, 3700 J of mechanical energy are somehow "lost" as the firefighter slides down the pole.

Core Concepts in College Physics Workbook

5. The fact that $E_1 \neq E_2$ does not violate the law of energy conservation. Mechanical (kinetic plus potential) energy remains unchanged only when no nonconservative forces perform work on the system. Work done by a nonconservative force (such as the friction $\vec{f}$ in this problem) can change mechanical energy into other forms, such as heat or sound waves. The 3700 J of energy "lost" by the firefighter does not vanish, but merely changes form, so the law of energy conservation still holds true.

6. By the work-energy theorem, the work done by the nonconservative force of friction is equal to the change in the system's mechanical energy

$$W_{n\text{-}c} = E_2 - E_1 = -3700 \text{ J}$$

7. When a constant force $\vec{F}$ acts on an object moving through a displacement $\vec{s}$, it performs work

$$W = \vec{F} \cdot \Delta\vec{s}$$

In this case, both the nonconservative force $\vec{f}$ and the displacement have only y components, so their dot product is simply

$$W_{n\text{-}c} = f\Delta y = f(y_2 - y_1)$$

Note that f is positive (because the force of friction on the firefighter is directed upward) while Δy is negative (because the firefighter's displacement is downward), so the work done by friction on the firefighter is negative (as you'd expect from the dot product of two vectors pointing in opposite directions). Physically, this represents the fact that friction reduces the system's total mechanical energy. If it had done positive work, it would increase the mechanical energy.

8. Equating the two expressions for $W_{n\text{-}c}$,

$$E_2 - E_1 = f(y_2 - y_1)$$

$$f = \frac{E_2 - E_1}{y_2 - y_1}$$

$$= \frac{-3700 \text{ J}}{-6.0 \text{ m}} = 617 \text{ N}$$

Thus, friction exerts an upward (positive) force of 617 N on the firefighter as he descends.

PROBLEM 31 Power

Before we begin...

1.

2. The given information is

$m = 650$ kg $\quad\quad v_0 = 0$

$v_f = 1.75$ m/s $\quad\quad \Delta t = 3.00$ s

3. Kinetic energy and gravitational potential energy will be changed by the work done by the elevator motor.

Solving the problem

4. The change in kinetic energy is

$$\Delta K = \tfrac{1}{2}mv_f^2 - \tfrac{1}{2}mv_0^2 = 995 \text{ J}$$

5. From kinematics, the relationship between height, time interval, and average velocity is

$$\Delta y = \langle v \rangle \Delta t$$

For constant acceleration,

$$\langle v \rangle = \frac{v_0 + v_f}{2}$$

Substituting,

$$\Delta y = \frac{(0 + 1.75 \text{ m/s})}{2}(3.00) = 2.63 \text{ m}$$

The change in potential energy is therefore

$$\Delta U = mg\Delta y = (650 \text{ kg})(9.80 \text{ m/s}^2)(2.63 \text{ m})$$
$$= 16{,}750 \text{ J}$$

The total work done by the motor in the first three seconds is 1.77×10^4 J.

(continued on next page . . .)

6. The average power is calculated as the work done by the motor divided by the time required:

$$P = \frac{\Delta U + \Delta K}{\Delta t} = \frac{1.77 \times 10^4 \text{ J}}{3 \text{ s}} = 5910 \text{ Watts}$$

7. Once the elevator is moving at a constant speed, the net force must be zero, so the force must equal the weight of the elevator.

Using $\vec{P} = \vec{F} \cdot \vec{v}$ after the elevator has reached its constant velocity,

$$P = (650 \text{ kg})(9.80 \text{ m/s}^2)(1.75 \text{ m/s})$$
$$= 1.11 \times 10^4 \text{ W}$$

PROBLEM 32 Power

Before we begin...

1. The relevant information is

$$m = 1500 \text{ kg} \qquad v_0 = 0$$
$$\Delta t = 3.00 \text{ s} \qquad v_f = 15.0 \text{ m/s}$$

2. The power exerted by any force is the rate of work per unit time performed by this force.

$$P = \frac{\Delta W}{\Delta t}$$

where ΔW denotes the amount of work performed in the time period Δt.

Solving the problem

3. Kinetic energy is the energy of motion. Since the car is initially not moving, its initial kinetic energy is

$$K_i = 0$$

In its final state, the car is moving at speed v_f, so its final kinetic energy is

$$K_f = \frac{1}{2} m v_f^2$$
$$= \frac{1}{2}(1500 \text{ kg})(15.0 \text{ m/s})^2$$
$$= 1.69 \times 10^5 \text{ J}$$

4. By the work-energy theorem, the net work done by all forces acting on the car must equal the change in its kinetic energy.

$$\Delta W_{net} = K_f - K_i$$

Because we have neglected friction and air resistance, all this work must have been supplied by the car's engine. Thus, the work ΔW done by the engine is

$$\Delta W = \frac{1}{2}mv_f^2 - 0 = 1.69 \times 10^5 \text{ J}$$

5. By the definition of power,

$$P = \frac{\Delta W}{\Delta t}$$

$$= \frac{1.69 \times 10^5 \text{ J}}{3.00 \text{ s}} = 5.63 \times 10^4 \text{ watts}$$

6. The work-energy theorem applies to work done by forces of all sorts, conservative and nonconservative. So for the car to reach the speed v_f, the net work done must be $\frac{1}{2}mv_f^2$, as we calculated before. The forces of friction and air resistance are directed opposite to the car's direction of motion, so any work they do must be negative. If these forces do significant work, the car's engine must do extra positive work in order to produce a total work of $\frac{1}{2}mv_f^2$. In order to do more work in the same amount of time, the engine must exert more power than it would if friction were negligible.

7. At an instant when the car is moving at velocity $\vec{v}$ while subjected to a force $\vec{F}$, the power delivered by this force is

$$P = \vec{F} \cdot \vec{v}$$

In this case, the force on the car points in the same direction as its velocity, so the vector dot product reduces to a simple numerical product

$$P = Fv$$

8. By the definition of acceleration,

$$a = \frac{\Delta v}{\Delta t}$$

$$= \frac{15.0 \text{ m/s}}{3.00 \text{ s}} = 5.00 \text{ m/s}^2$$

9. By Newton's second law, the net force acting on the car is given by

$$F = ma$$

$$= (1500 \text{ kg})(5.00 \text{ m/s}^2) = 7500 \text{ N}$$

Also, since the car's acceleration is constant, we can find its speed after $t = 2.0$ s simply by

$$v = v_0 + at$$
$$= 0 + (5.00 \text{ m/s}^2)(2.0 \text{ s}) = 10.0 \text{ m/s}$$

10. Substituting the results of step 9 into the equation cited in step 7, we find

$$P = Fv$$
$$= (7500 \text{ N})(10.0 \text{ m/s}) = 7.50 \times 10^4 \text{ W}$$

Note that this is more than the average power delivered over the entire three seconds. The force on the car is constant, so the engine delivers less power near the beginning of the interval, when the car's speed v is lower, and more near the end, when v is higher.

PROBLEM 33 Conservation of Energy

Before we begin...

1. The law of conservation of energy states that the total mechanical energy of a system (kinetic energy + potential energy) must remain constant in any isolated system of objects that interact only through conservative forces.

2. The given information is

$$K_i = 30 \text{ J} \qquad \Sigma U_i = 10 \text{ J} \qquad K_f = 18 \text{ J}$$

Solving the problem

3. The total energy E_i at time t_i is

$$E_i = K_i + U_i = 40 \text{ J}$$

4. Applying the law of conservation of energy for a conservative system, the total energy must remain constant.

$$E_f = E_i = 40 \text{ J}$$

Expressing the total final energy in terms of the kinetic and potential energies, we can solve for the unknown, potential energy

$$E_f = K_f + U_f$$

$$U_f = E_f - K_f = 40 \text{ J} - 18 \text{ J} = 22 \text{ J}$$

5. If $U_f = 5$ J, then the final mechanical energy

$$E_f = K_f + U_f = 18 \text{ J} + 5 \text{ J} = 23 \text{ J}$$

The difference $E_f - E_t = -17$ J is the result of work done by nonconservative forces on the particle.

Module 6 Momentum

PROBLEM 34 Conservation of Momentum

Before we begin...

1. One suitable reference frame has as origin the astronaut's initial position, with the positive direction considered the same as the direction the astronaut throws her oxygen bottle.

2. The given information is

$$m_{astronaut} = 60.0 \text{ kg} \qquad m_{tank} = 10.0 \text{ kg}$$
$$v_{tank\text{-}initial} = 0.0 \text{ m/s} \qquad v_{tank\text{-}final} = 12.0 \text{ m/s}$$
$$v_{astronaut\text{-}initial} = 0.0 \text{ m/s} \qquad \Delta t = 60.0 \text{ s}$$

3. The astronaut's velocity and the time interval are used to calculate the distance traveled by

$$d = v_{astronaut\text{-}final} \, \Delta t$$

4. Newton's second law expressed in terms of momentum is

$$F_{net} = \frac{\text{change in momentum}}{\text{time interval}} = \frac{\Delta p}{\Delta t}$$

5. The astronaut throws her oxygen tank by exerting a force on it. According to Newton's third law, an equal and opposite force is applied to the astronaut. Since there are no external forces acting on them, the net force on the system of the astronaut and the tank must be

$$F_{net} = 0$$

6. With no net force on the system, the total momentum of the system must be constant. Thus, momentum is conserved. This is expressed by

$$\Delta p = p_{final} - p_{initial} = 0$$

Solving the problem

7. The initial and final momentums of the system are

$$p_{initial} = m_{astronaut} v_{astronaut\text{-}initial} + m_{tank} v_{tank\text{-}initial}$$
$$p_{final} = m_{astronaut} v_{astronaut\text{-}final} + m_{tank} v_{tank\text{-}final}$$

8. Substituting these momentum expressions into the conservation equation, and recognizing that the initial momentum is zero,

$$p_{final} = m_{astronaut} v_{astronaut\text{-}final} + m_{tank} v_{tank\text{-}final} = 0$$

Solving for $v_{astronaut\text{-}final}$,

$$v_{astronaut\text{-}final} = \frac{-(m_{tank} v_{tank\text{-}final})}{m_{astronaut}}$$
$$= \frac{-(10.0 \text{ kg})(12.0 \text{ m/s})}{60.0 \text{ kg}}$$
$$= -2.0 \text{ m/s}$$

9. Using the equation from step 3, the above expression and substituting the known values,

$$d = v_{astronaut\text{-}final} \Delta t$$
$$= (-2.0 \text{ m/s})(60.0 \text{ s})$$
$$= -120 \text{ m}$$

The astronaut can move 120 meters toward the shuttle in 60.0 seconds.

PROBLEM 35 Impulse

Before we begin...

1. The given information is

$$m = 2.0 \text{ kg}$$
$$v_{i(\text{particle initially at rest})} = 0$$
$$v_{i(\text{particle initially moving})} = -2.0 \text{ m/s}$$

2. Impulse is related to force and the time interval over which the force is applied by the equation

$$\Delta \vec{p} = \vec{F} \Delta t$$

3. Momentum is defined as $\vec{p} = m\vec{v}$.

Solving the problem

4. The three regions are two triangles and one rectangle with respective areas of

4 N·s, 4 N·s, and 4 N·s

The total impulse therefore is 12 N·s.

5. Final momentum can be expressed in terms of the initial momentum and the impulse as

$$\vec{p}_f = \vec{p}_i + \Delta \vec{p}$$

6. For $\vec{v}_i = 0$,

$$\vec{p}_f = 0 + 12 \text{ N·s} = 12 \text{ N·s}$$

Because $\vec{p} = m\vec{v}$,

$$\vec{v}_f = 6.0 \text{ m/s}$$

If $\vec{v}_1 = -2.0 \text{ m/s}$,

$$\vec{p}_f = -4 \text{ N·s} + 12 \text{ N·s} = 8 \text{ N·s}$$

Then,

$$\vec{v}_f = 4.0 \text{ m/s}$$

7. A constant force required to give the same value of impulse is

$$12.0 \text{ N·s} = \langle \vec{F} \rangle (5.0 \text{ s})$$

so

$$\langle \vec{F} \rangle = 2.40 \text{ N}$$

PROBLEM 36 Perfectly Inelastic Collisions

Before we begin...

1.

2. The given information is

$m_1 = 90$ kg $\vec{v}_1 = +10$ m/s
$m_2 = 120$ kg $\vec{v}_2 = -4.0$ m/s

3. The two masses will stick and move as one.

4. Momentum is always conserved in a collision in an isolated system.

Solving the problem

5. Evaluating the total momentum of the system before the collision, we find

$\vec{p}_T = \vec{p}_1 + \vec{p}_2$
$= (90 \text{ kg})(+10 \text{ m/s}) + (120 \text{ kg})(-4.0 \text{ m/s})$
$= +420$ N·s.

6. Because the momentum after the collision is the same as before the collision, the velocity of the players (who stick together and move as one mass) can be evaluated as

$\vec{p} = (m_1 + m_2)\vec{v}_c$

$\vec{v}_c = \dfrac{+420 \text{ N·s}}{210 \text{ kg}} = +2.0$ m/s

Because the + direction was selected as the direction of the halfback's velocity (north) and because the final velocity is also positive, the players are moving north after the collision.

7. Before the collision, the total kinetic energy was

$K_1 + K_2 = \frac{1}{2}m_1v_1^2 + \frac{1}{2}m_2v_2^2 = 5460$ J

After the collision, it becomes

$K_C = \frac{1}{2}(m_1 + m_2)v_c^2 = 420$ J

The work that the players did on each other during the collision converted much of their kinetic energy into other forms. Most of this energy ends up as heat, while smaller amounts end up as sound waves and as lingering vibrations in the players' helmets or skeletal systems.

PROBLEM 37 Perfectly Inelastic Collisions

Before we begin...

1.

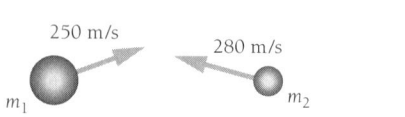

2. The given information is

$m_1 = 5$ g $\vec{v}_1 = (250$ m/s, $20°)$
$m_2 = 3$ g $\vec{v}_2 = (280$ m/s, $165°)$

3. The two bullets undergo a perfectly inelastic collision.

4. The problem must be solved in two dimensions because the motion is in two dimensions.

Solving the problem

5. The x and y components of momentum for each bullet before the collision are

$p_{1x} = m_1v_1 \cos\theta_1 = 1175$ g·m/s

$p_{1y} = m_1v_1 \sin\theta_1 = 428$ g·m/s

$p_{2x} = m_2v_2 \cos\theta_2 = -811$ g·m/s

$p_{2y} = m_2v_2 \sin\theta_2 = 217$ g·m/s

6. Solving for v_x,

$(m_1 + m_2)v_x = p_{1x} = p_{2x}$

$v_x = \dfrac{(p_{1x} + p_{2x})}{(m_1 + m_2)} = \dfrac{364 \text{ g·m/s}}{8 \text{ g}} = 45.5$ m/s

7. For the y component we find

$v_y = \dfrac{(p_{1x} + p_{2x})}{(m_1 + m_2)} = \dfrac{645 \text{ g·m/s}}{8 \text{ g}} = 80.6$ m/s

8. Evaluating to find the velocity of the combined mass after the collision, we find

$v_c = \sqrt{v_x^2 + v_y^2} = 92.5$ m/s

$\theta = \tan^{-1}\left(\dfrac{v_y}{v_x}\right) = 60.6°$

Core Concepts in College Physics Workbook

PROBLEM 38 Perfectly Elastic Collisions

Before we begin...

1. The given information is

$m_1 = 2.00$ kg $m_2 = 4.00$ kg $\Delta h = -5$ m

$v_1 = 0$ $v_2 = 0$

2. Potential energy is converted into kinetic energy.

3. Linear momentum and kinetic energy are both conserved in an elastic collision.

Solving the problem

4. Using the law of conservation of energy to solve for the speeds of the two objects immediately before the collision,

$\Delta U + \Delta K = 0$ for each object

(No nonconservative forces are present.)

$mg\Delta h + (\frac{1}{2}mv^2 - 0) = 0$

The mass cancels in the equation, so solving for the speed of each object gives

$v = \sqrt{2gh} = 9.90$ m/s

From this, we conclude $v_1 = +9.90$ m/s and $v_2 = -9.90$ m/s immediately before the collision.

5. The total linear momentum and kinetic energy are

$p_T = (2.00$ kg$)(+9.90$ m/s$)$
$\quad + (4.00$ kg$)(-9.90$ m/s$) = -19.8$ N·s

$K_T = \frac{1}{2} (2.00$ kg$)(9.90$ m/s$)^2$
$\quad + \frac{1}{2} (4.00$ kg$)(9.90$ m/s$)^2 = 294$ J

Hence, after the collision,

$2v_1 + 4v_2 = -19.8$ and

$\frac{1}{2}(2)v_1^2 + \frac{1}{2}(4)v_2^2 = 294$

(The units have been dropped, but will yield velocities in m/s.)

Solving for v_1 yields two possible solutions:

$v_1 = +9.90$ m/s   and $v_1 = -16.5$ m/s

The first solution is rejected; it describes the condition before the collision. The appropriate solution is $v_1 = -16.5$ m/s.

Substitution back into the momentum equation with the value of v_1 yields

$v_2 = +3.3$ m/s

6. The law of conservation of energy for Block 1 can be expressed as

$(0 - \frac{1}{2}m_1v_1^2) + (m_1gh_1 - 0) = 0$

When we substitute the speed of the object after the collision, the height is

$h_1 = \dfrac{v_1^2}{2g} = \dfrac{(-16.9 \text{ m/s})^2}{2(9.8 \text{ m/s}^2)} = 13.9$ m

Likewise, for object m_2, we obtain

$h_2 = 0.56$ m

PROBLEM 39 Perfectly Elastic Collision

Before we begin...

1. Both kinetic energy and momentum are conserved in an elastic collision.

2. Conservation of energy is expressed as

$$mv^2_{\text{orange-initial}} = mv^2_{\text{orange-final}} + mv^2_{\text{yellow-final}}$$

Conservation of momentum is expressed

$$m\vec{v}_{\text{orange-initial}} = m\vec{v}_{\text{orange-final}} + m\vec{v}_{\text{yellow-final}}$$

In component form this is

$$mv_{\text{orange-initial-}x} = mv_{\text{orange-final-}x} + mv_{\text{yellow-final-}x}$$

$$mv_{\text{orange-initial-}y} = mv_{\text{orange-final-}y} + mv_{\text{yellow-final-}y}$$

Since the masses of the disk are equal, we can divide all these equations by m.

3.

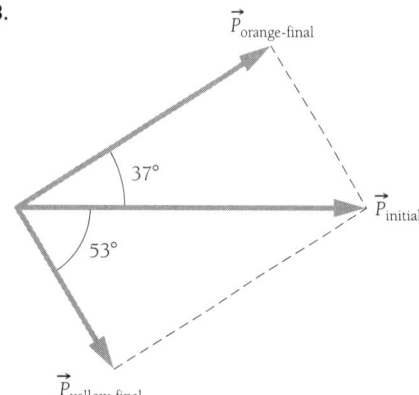

4. We choose the x axis along the initial direction of motion of the orange disk. The velocity components are

$$v_{\text{orange-initial-}x} = v_{\text{orange-initial}} = 5.0 \text{ m/s}$$
$$v_{\text{orange-initial-}y} = 0.0 \text{ m/s}$$
$$v_{\text{orange-final-}x} = v_{\text{orange-final}} \cos 37°$$
$$v_{\text{orange-final-}y} = v_{\text{orange-final}} \sin 37°$$
$$v_{\text{yellow-final-}x} = v_{\text{yellow-final}} \cos 53°$$
$$v_{\text{yellow-final-}y} = -v_{\text{yellow-final}} \sin 53°$$

Solving the problem

5. Solving the y component momentum equation and substituting the known values,

$$0 = v_{\text{orange-final-}y} + v_{\text{yellow-final-}y}$$
$$0 = v_{\text{orange-final}} \sin 37° - v_{\text{yellow-final}} \sin 53°$$

$$v_{\text{orange-final}} = \frac{v_{\text{yellow-final}} \sin 53°}{\sin 37°}$$

We can use trigonometry to express

$$-\sin 53° = -\cos(90° - 53°) = -\cos 37°$$
$$\cos 53° = \sin(90° - 53°) = \sin 37°$$

Substituting in the y component momentum equation from above,

$$v_{\text{orange-final}} = \frac{v_{\text{yellow-final}} \cos 37°}{\sin 37°}$$

6. Using the x component momentum equation from step 2, the velocity components from step 4, and the second trigonometric identity above, yields

$$v_{\text{orange-initial}} = v_{\text{orange-final}} \cos 37° + v_{\text{yellow-final}} \sin 37°$$

Substituting from the expression for $v_{\text{orange-final}}$ from above,

$$v_{\text{orange-initial}}$$
$$= \frac{v_{\text{yellow-final}} \cos^2 37°}{\sin 37°} + v_{\text{yellow-final}} \sin 37°$$
$$= \frac{(\cos^2 37° + \sin^2 37°) v_{\text{yellow-final}}}{\sin 37°}$$

Rearranging

$$v_{\text{yellow-final}} = v_{\text{orange-initial}} \sin 37° = 3.0 \text{ m/s}$$

Substituting this into the last equation in step 5 yields

$$v_{\text{orange-final}} = v_{\text{orange-initial}} \cos 37° = 4.0 \text{ m/s}$$

Following up

Note that the energy conservation equation was not explicitly used to solve the problem. If the direction of the yellow disk relative to the orange disk had not been specified, the energy equation would have been necessary. This direction is not happenstance, but is a consequence of energy conservation. By drawing a vector diagram of the momenta, relating the initial momentum of the orange disk, to the final momenta of both disks, it becomes apparent that the energy equation, in the case where the disks have equal mass, is a statement of the Pythagorean Theorem for Right Triangles: the sum of the squares of the two sides is equal to the square of the hypotenuse, or, the sum of the square of the final velocities is equal to the square of the initial velocity.

In any elastic collision between two objects of equal mass, their final velocities will be at right angles, when viewed from the reference frame of one of the objects.

PROBLEM 40 Center of Mass

Before we begin...

1. Equal mass is located above and below the x axis at the same distance from the axis.

2. Based on symmetry, the center of mass will lie on the x axis; $y_{CM} = 0$.

Solving the problem

3. Resolving the distance of the hydrogen atoms to the oxygen atom, we find the x component

$$x = L \cos \theta = (0.100 \text{ nm}) \cos 53°$$
$$= 0.0602 \text{ nm}$$

4. Using the equation for the location of the x_{CM} for a set of discrete particles, we find

$$x_{CM} = \frac{\sum m_i x_i}{\sum m_i}$$

$$= \frac{(15.99 \text{ u})(0) + (1.008 \text{ u})(0.0602 \text{ nm}) + (1.008 \text{ u})(0.0602 \text{ nm})}{(15.999 + 1.008 + 1.008) \text{ u}}$$

$$= 0.00673 \text{ nm}$$

The coordinates of the center of mass are (0.00673, 0) nm from the oxygen nucleus.

Module 7 Rotational Mechanics

PROBLEM 41 Rotational Kinematics

Before we begin...

1. The relevant information is
$$\omega_0 = 0$$

$$\omega_f = \left(\frac{2.51 \times 10^4 \text{ rev}}{\text{min}}\right)\left(\frac{2\pi \text{ rad}}{\text{rev}}\right)\left(\frac{1 \text{ min}}{60 \text{ s}}\right)$$

$$= 2630 \text{ rad/s}$$

$$\Delta t = 3.20 \text{ s}$$

2. This problem is analogous to one in which an object starts from rest and moves in a straight line under constant acceleration, reaching a known final velocity in a known amount of time. Any techniques that would find the acceleration and distance traveled in such a linear-motion problem translate directly into methods for finding the angular acceleration and the angle turned through in this problem.

Solving the problem

3. Since we know the initial and final values of angular velocity ω_0 and ω_f, and the elapsed time Δt, the most appropriate kinematic equation is

$$\alpha = \frac{\Delta\omega}{\Delta t} = \frac{\omega_f - \omega_0}{\Delta t}$$

4. Substituting known values into the above equation,

$$\alpha = \frac{2630 \text{ rad/s}}{3.20 \text{ s}} = 822 \text{ rad/s}^2$$

Note that the units of radians per second squared are the expected ones for an angular acceleration.

5. We can find the angle $\Delta\theta$ that the drill turns through by

$$\Delta\theta = \omega_0 t + \frac{1}{2}\alpha t^2$$

Since the drill's initial angular velocity ω_0 is zero, this reduces to

$$\Delta\theta = \frac{1}{2}\alpha t^2$$

6. Substituting known values,

$$\Delta\theta = \frac{1}{2}(822 \text{ rad/s}^2)(3.20 \text{ s})^2$$

$$\Delta\theta = 4210 \text{ rad}$$

7. Converting radians into revolutions, we find

$$(4210 \text{ rad})\left(\frac{1 \text{ revolution}}{2\pi \text{ rad}}\right) = 670 \text{ revolutions}$$

Thus, the drill turns 670 times in the 3.2-second interval while it is accelerating to full speed.

Core Concepts in College Physics Workbook

PROBLEM 42 Rotational Kinematics

Before we begin...

1. The relevant information is

$v_0 = 0$

$v_f = 25.0$ m/s

$\Delta\theta = (1.25 \text{ rev})\left(\dfrac{2\pi \text{ rad}}{\text{rev}}\right) = 7.85$ rad

$r = 1.10$ m

2. When an object moves around a circular path of radius r, its angular speed ω and its linear speed v are related by

$$v = r\omega, \quad \text{or} \quad \omega = \dfrac{v}{r}$$

3. Applying the above relation to the initial and final speeds of the discus,

$$\omega_0 = \dfrac{v_0}{r} = \dfrac{0}{1.10 \text{ m}} = 0$$

$$\omega_f = \dfrac{v_f}{r} = \dfrac{25.0 \text{ m/s}}{1.10 \text{ m}} = 22.7 \text{ rad/s}$$

(*Note:* It is disconcerting that units of radians often seem to "come out of nowhere" as in the above calculation for ω_f. This can happen because radians are actually a dimensionless unit. That is, an angle in radians is the ratio of an arc length to a radius, so radians are equivalent to length over length, which is dimensionless.)

Solving the problem

4. Since we already know the initial and final angular speeds ω_0 and ω_f and the angle through which the discus rotates $\Delta\theta$, the kinematic equation best suited to finding α is

$$\omega_f^2 = \omega_0^2 + 2\alpha\Delta\theta$$

In this case, since $\omega_0 = 0$, this reduces to

$$\omega_f^2 = 2\alpha\Delta\theta$$

5. Solving for α,

$$\alpha = \dfrac{\omega_f^2}{2\Delta\theta}$$

Note that since ω_f has units of rad/s, and $\Delta\theta$ has units of radians, the expression for α yields units of

$$\dfrac{(\text{rad/s})^2}{\text{rad}} = \dfrac{\text{rad}^2/\text{s}^2}{\text{rad}} = \text{rad/s}^2$$

which are the correct units for an angular acceleration.

6. Substituting known values,

$$\alpha = \dfrac{(22.7 \text{ rad/s})^2}{2(7.85 \text{ rad})} = 32.8 \text{ rad/s}^2$$

7. Knowing the angular acceleration α, there are many ways to find the time t. The simplest is by using

$$\omega_f = \omega_0 + \alpha t$$

Which reduces in this case, where $\omega_0 = 0$, to

$$\omega_f = \alpha t$$

8. Solving for t,

$$t = \dfrac{\omega_f}{\alpha}$$

Note that since ω_f has units of rad/s, and α has units of rad/s^2, the expression for t yields units of

$$\dfrac{\text{rad/s}}{\text{rad/s}^2} = \text{s}$$

which are the correct units for time.

9. Substituting known values,

$$t = \dfrac{22.7 \text{ rad/s}}{32.8 \text{ rad/s}^2} = 0.69 \text{ s}$$

PROBLEM 43 Rotational Kinetic Energy

Before we begin...

1. $R = 0.80$ m

   $m = 500$ kg

   $\omega_0 = \left(\dfrac{8000 \text{ rev}}{\text{min}}\right)\left(\dfrac{2\pi \text{ rad}}{\text{rev}}\right)\left(\dfrac{1 \text{ min}}{60 \text{ s}}\right)$

   $= 838$ rad/s

   $f = 1000$ N

   $d = 5.00 \times 10^4$ m

2. The moment of inertia of a uniform cylinder of mass m and radius R is given by

   $I = \dfrac{1}{2} mR^2$

   Thus, the flywheel's moment of inertia is

   $I = \dfrac{1}{2}(500 \text{ kg})(0.80 \text{ m})^2$

   $I = 160$ kg·m^2

Solving the problem

3. The rotational kinetic energy of an object of moment of inertia I rotating with angular speed ω is given by

   $K = \dfrac{1}{2} I \omega^2$

   For the flywheel's given moment of inertia and initial angular speed, this gives an initial kinetic energy of

   $K_i = \dfrac{1}{2}(160 \text{ kg m}^2)(838 \text{ rad/s})^2$

   $K_i = 5.62 \times 10^7$ J

4. Since the force of air resistance $\vec{f}$ is directed exactly opposite the car's direction of motion, the work done will be negative force times distance:

   $W = -fd$

   By the work-energy theorem, this negative work will decrease the car's total energy by an amount fd. The car draws its energy from the flywheel, so if the wheel begins with initial kinetic energy $\frac{1}{2} I \omega_0^2$, its energy after traveling a distance d is

   $K = \dfrac{1}{2} I \omega_0^2 - fd$

5. Combining the relation just derived with the general expression $K = \frac{1}{2} I \omega^2$, we can find the flywheel's angular speed ω after the car travels a distance d:

   $\dfrac{1}{2} I \omega^2 = \dfrac{1}{2} I \omega_0^2 - fd$

   $\omega^2 = \omega_0^2 - \dfrac{2fd}{I}$

   $\omega = \sqrt{\omega_0^2 - \dfrac{2fd}{I}}$

6. Substituting known values into this expression,

   $\omega = \sqrt{(838 \text{ rad/s})^2 - \dfrac{2(1000 \text{ N})(5.0 \times 10^4 \text{ m})}{160 \text{ kg m}^2}}$

   $= 278$ rad/s

7. Our expression for the flywheel's angular speed ω should yield 0 when the car has traveled its maximum possible distance d_{max}. In other words,

   $0 = \sqrt{\omega_0^2 - \dfrac{2f d_{max}}{I}}$

   $0 = \omega_0^2 - \dfrac{2f d_{max}}{I}$

   $d_{max} = \dfrac{I \omega_0^2}{2f}$

8. The fastest way to verify the dimensions of our expression for d_{max} is to note that $I\omega_0^2$ is a familiar expression for energy, which has units of Joules, or Newton-meters. The force f has units of Newtons. So the expression has units of Newton-meters over Newtons, or simply meters; these are the correct units for a distance.

It's also important to match the equation against your physical intuition. If the flywheel had a higher initial rotation rate ω_0, or if a heavier flywheel with a higher moment of inertia I were given the same initial rotation rate, then the wheel would start with more kinetic energy and the car should travel farther. Thus, both I and ω_0 should appear in the numerator of the equation for d_{max}. On the other hand, if the air resistance f were stronger, then the car would not travel as far, so f should appear in the denominator of the expression for d_{max}. All of these intuitive claims are borne out in the expression we've derived.

9. Substituting known values,

$$d_{max} = \frac{(160 \text{ kg m}^2)(838 \text{ rad/s})^2}{2(1000 \text{ N})}$$
$$= 5.62 \times 10^4 \text{ m}$$
$$= 56.2 \text{ km}$$

PROBLEM 44 Torque

Before we begin...

1. $R = 2.0$ m
   $m = 3.0 \times 10^5$ kg
   $\omega_0 = 0$ (Satellite is initially at rest.)
   $f = 2500$ N
   $t = 4$ min $= 240$ s

2. The moment of inertia of a uniform sphere of mass m and radius R is given by

$$I = \frac{2}{5}mR^2$$

Thus, the satellite's moment of inertia is

$$I = \frac{2}{5}(3.0 \times 10^5 \text{ kg})(2.0 \text{ m})^2$$

$$I = 4.8 \times 10^5 \text{ kg m}^2$$

Solving the problem

3. The forces exerted by the four engines are equally strong and all point in different directions:

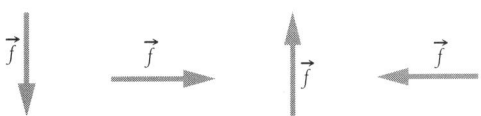

The sum of these four vectors is zero because the upward-pointing vector cancels the downward-pointing one, and the leftward one cancels the rightward one.

Since the net force on the satellite is zero, it undergoes no linear acceleration—that is, its center of mass remains at rest, although it will quite obviously begin to rotate.

(continued on next page . . .)

4. Each engine exerts a force f on a point a distance r from the satellite's axis of rotation. The direction of each engine's force makes an angle $\theta = 90°$ with the line running from the axis to the engine. Thus, each engine exerts a torque

$$\tau = rf \sin \theta$$
$$= (2.0 \text{ m})(2500 \text{ N}) \sin 90°$$
$$= 5000 \text{ N·m}$$

about the axis of rotation.

All four of these torques act in the same direction (they all tend to turn the satellite counterclockwise as shown in the diagram).

5. Since all four torques act in the same direction, their magnitudes can simply be added together; the net torque on the satellite has magnitude

$$\tau_{net} = 20{,}000 \text{ N·m}$$

6. Newton's second law for rotational motion relates an object's angular acceleration α to the torque τ_{net} acting on it and its moment of inertia I.

$$\alpha = \frac{\tau_{net}}{I}$$

In this case, it yields an angular acceleration

$$\alpha = \frac{20000 \text{ N·m}}{4.8 \times 10^5 \text{ kg·m}^2}$$
$$= 0.042 \text{ rad/s}^2$$

7. The satellite's final angular speed ω can be found using the kinematic equation

$$\omega = \omega_0 + \alpha t$$

Substituting known values into this equation,

$$\omega = 0 + (0.042 \text{ rad/s}^2)(240 \text{ s})$$
$$= 10.0 \text{ rad/s}$$

8. The angle θ that the satellite turns through can be found using any of several methods; one simple one is

$$\Delta\theta = \omega_0 t + \frac{1}{2}\alpha t^2$$

Substituting known values into this equation,

$$\Delta\theta = 0 + \frac{1}{2}(0.042 \text{ rad/s})(240 \text{ s})^2$$
$$= 1200 \text{ rad}$$

Because the problem asked for this angle in revolutions, rather than radians, we simply convert it:

$$\Delta\theta = 1200 \text{ rad}\left(\frac{1 \text{ rot}}{2\pi \text{ rad}}\right) = 191 \text{ rot}$$

9. The satellite's kinetic energy could be calculated directly using

$$K_R = \frac{1}{2}I\omega^2$$

Another method could be to use the rotational work-energy theorem. Because the initial kinetic energy is zero, the final energy is simply equal to the work done:

$$K_R = \tau_{net} \Delta\theta$$

Or, since each engine exerts a linear force f along a path of length $s = r\Delta\theta$, we can calculate the work done by each engine using $W = fs = fr\Delta\theta$. Because there are four engines, the final energy is

$$K_R = 4fr\Delta\theta$$

Naturally, none of these methods will "yield the largest answer for the energy." Because all three are correct methods, all three will yield the same answer—namely, 2.4×10^7 J.

Core Concepts in College Physics Workbook

PROBLEM 45 Work and Energy in Rotational Motion

Before we begin...

1. The given information is

   $M = 100$ kg $R = 0.5$ m

   $\Delta t = 6.0$ s $\omega_i = 50$ rev/min

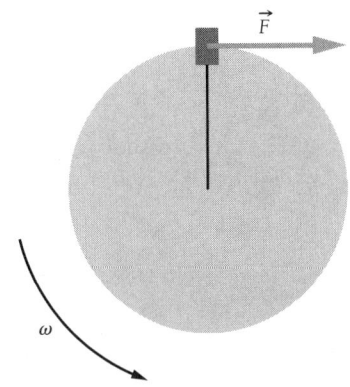

2. The frictional torque opposes the direction of rotation and thus decreases the rotational speed of the wheel.

3. The final rotational kinetic energy of the wheel will be zero.

Solving the problem

4. Using the equation $I = \frac{1}{2}MR^2$ for a solid disk, we calculate the moment of inertia of the wheel to be

   $I = \frac{1}{2}(100 \text{ kg})(0.5 \text{ m})^2 = 12.5$ kg·m²

5. The angular speed is converted to

   $\omega_i = (50 \text{ rev/min})(2\pi \text{ rad/rev})(1 \text{ min}/60 \text{ s})$
   $= 5.23$ rad/s

   This must be done so that the expression will be in fundamental units.

6. The initial rotational kinetic energy and the change in kinetic energy during the problem are

   $K_i = \frac{1}{2}I\omega_i^2 = 171$ J

   $\Delta K = (0 - 171 \text{ J}) = -171$ J

7. The formula for the work done by a constant net torque is

   $\tau_{net}\Delta\theta = \Delta K$

8. To find $\Delta\theta$,

   $\dfrac{\omega_i + \omega_f}{2} = \dfrac{\Delta\theta}{\Delta t}$ so $\Delta\theta = 15.7$ rad

9. Using the work-energy theorem to solve for the torque,

   $(\tau_{net})(15.7 \text{ rad}) = 171$ J

   $\tau_{net} = 10.9$ N·m

   Note that the negative sign indicates the direction of the torque and is therefore ignored in computing the magnitude.

10. The frictional force $\vec{F}$ acts at a right angle to the line from the wheel's axis to the point where the force is applied. Thus, the torque exerted by this force about the axis is simply

    $\tau = RF$

    Or, solving for F,

    $F = \dfrac{\tau}{R} = \dfrac{10.9 \text{ N·m}}{0.5 \text{ m}}$

    $F = 21.8$ N

PROBLEM 46 Rolling Motion

Before we begin...

1.

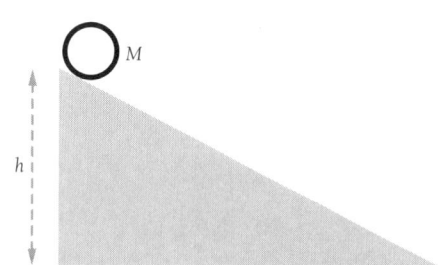

2. Each object has potential energy at the top of the incline.

Solving the problem

3. A rolling object has both rotational kinetic energy and translational kinetic energy. The expressions of the two types of kinetic energy are

$$K_{rot} = \frac{1}{2} I\omega^2 \qquad K_{tran} = \frac{1}{2} Mv^2$$

4. Since no nonconservative forces do work on the rolling objects, the law of energy conservation states that for each object,

$$U_i + K_{i\text{-rot}} + K_{i\text{-tran}} = U_f + K_{f\text{-rot}} + K_{f\text{-tran}}$$

The initial kinetic energies $K_{i\text{-rot}}$ and $K_{i\text{-tran}}$ and the final gravitational potential energy U_f are all zero, so this equation reduces to

$$U_i = K_{f\text{-rot}} + K_{f\text{-tran}}$$

$$Mgh = \frac{1}{2} I\omega^2 + \frac{1}{2} Mv^2$$

5. For an object which rolls without slipping, $\omega = v/R$. Substituting this relation for ω and solving for v,

$$2Mgh = (I/R^2)v^2 + Mv^2$$

$$v = \sqrt{\frac{2gh}{1 + (I/MR^2)}}$$

6. For a disk, $(I/MR^2) = 1/2$, so

$$v_{disk} = \sqrt{\frac{4gh}{3}}$$

For a hoop, $(I/MR^2) = 1$, so

$$v_{hoop} = \sqrt{gh}$$

The disk has the greater speed at the bottom. The disk stores less of its kinetic energy in rotation, leaving more for translational kinetic energy.

7. Because the objects cover the same distance, the one with the greater average speed will reach the bottom first. Because they have the same initial speed, the one with the greater final speed will have the greater average speed as well.

Core Concepts in College Physics Workbook

PROBLEM 47 Conservation of Angular Momentum

Before we begin...

1. The law of conservation of angular momentum states that in the absence of a net external torque, the angular momentum of a system will remain unchanged. The angular momentum at any time will be equal to the angular momentum at any other time.

2. Angular momentum is related to the moment of inertia of a system by

$$L = I\omega$$

3. The given information is

$$I_{m\text{-}g\text{-}r} = 250 \text{ kg·m}^2$$

$$\omega_{m\text{-}g\text{-}r} = 10 \text{ rev/min} = 1.05 \text{ rad/s}$$

$$R_{m\text{-}g\text{-}r} = 2.0 \text{ m}$$

$$m_c = 25 \text{ kg}$$

4. The equation for computing the moment of inertia of a revolving point mass particle is

$$I = mR^2$$

Solving the problem

5. The angular momentum of the merry-go-round is expressed as

$$L = I\omega = (250 \text{ kg·m}^2)(1.05 \text{ rad/s}) = 262 \text{ J·s}$$

6. After the child jumps onto the merry-go-round, the new moment of inertia of the system is

$$I = I_{m\text{-}g\text{-}r} + I_c = 250 \text{ kg·m}^2 + (25 \text{ kg})(2.0 \text{ m})^2$$
$$= 350 \text{ kg·m}^2$$

7. Equating the initial and final angular momenta, substituting known quantities, and solving for ω_2,

$$\omega_2 = (0.75 \text{ rad/s})\left(\frac{1 \text{ rev}}{2\pi \text{ rad}}\right)(60 \text{ s/min})$$
$$= 7.16 \text{ rev/min}$$

Module 8 Simple Harmonic Motion

PROBLEM 48 Simple Harmonic Motion

Before we begin...

1. The mass of the particle does not matter because the equation of motion for the particle is independent of the mass.

2. The given information is

$$f = 3.0 \text{ Hz} \qquad A = 5.0 \text{ cm}$$

3.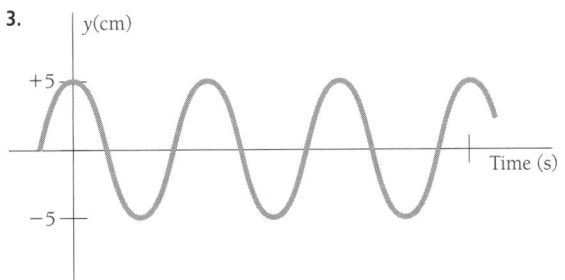

Solving the problem

4. The distance between the equilibrium point and the maximum displacement is 5.00 cm, therefore the total distance traveled in one cycle is 20 cm.

5. The expression for $v(t)$ is

$$v = -\omega A \sin(\omega t + \phi), \text{ where } \omega = 2\pi f$$

The maximum value of v is ωA, or

$$(2\pi)(3.0 \text{ Hz})(5.0 \text{ cm}) = 94.2 \text{ cm/s}$$

This occurs when $y = 0$.

6. The expression for $a(t)$ is

$$a = -\omega^2 A \cos(\omega t + \phi)$$

The maximum magnitude for the acceleration is $\omega^2 A$. This occurs when the displacement is a maximum, $y = \pm A$, $a_{\max} = (6\pi \text{ Hz})^2(5.0 \text{ cm}) = 1776.5 \text{ cm/s}^2$.

Module 8 **Simple Harmonic Motion**

PROBLEM 49 Physical Nature of Waves

Before we begin...

1. The general expression for a wave traveling to the left on a string is

$$y(x, t) = A \cos (kx + \omega t + \phi)$$

2. The transverse speed v_y and acceleration a_y on such a traveling wave are given by

$$v_y(x, t) = -\omega A \sin (kx + \omega t + \phi)$$

$$a_y(x, t) = -\omega^2 A \cos (kx + \omega t + \phi)$$

Solving the problem

3. Comparing the general expression with the given equation

$$A \cos (kx + \omega t + \phi) = (0.12 \text{ m}) \sin \pi[(x/8) + 4t]$$

allows the identification of the parameters

$A = 0.12$ m $\qquad \omega = 4\pi$

$\phi = -\pi/2$ $\qquad k = \pi/8$

Since $k = 2\pi/\lambda = \pi/8$ and $\omega = 2\pi/T = 4\pi$, comparison of terms yields

$\lambda = 16$ m $\qquad T = 0.5$ s

$v = \lambda/T = 32.0$ m/s

Remember that v, the wave speed, is the speed of the disturbance along the string and is distinct from v_y, the transverse speed of the string.

4. Using the values found for the various parameters in the expressions for transverse velocity and acceleration results in

$$v_y(x, t) = -4\pi(0.12 \text{ m}) \sin \pi\left(\frac{x}{8} + 4t - \frac{1}{2}\right)$$

$$a_y(x, t) = -16\pi^2(0.12 \text{ m}) \cos \pi\left(\frac{x}{8} + 4t - \frac{1}{2}\right)$$

For $x = 1.6$ m, $t = 0.2$ s,

$$v_y(1.6 \text{ m}, 0.2 \text{ s}) = -4\pi(0.12 \text{ m}) \sin \frac{\pi}{2}$$

$$= -1.51 \text{ m/s}$$

Core Concepts in College Physics Workbook

PROBLEM 50 Frequency, Wavelength, and Wave Speed

Before we begin...

1. The speed of a wave v is related to the wavelength and frequency by

$$v = f\lambda$$

2. The known values are

$$f = 60.0 \text{ kHz} = 60.0 \times 10^3 \text{ sec}^{-1}$$
$$v = 340 \text{ m/s}$$

Solving the problem

3. Solving the equation from step 1 for λ,

$$\lambda = \frac{v}{f}$$

4. Substituting the known values gives

$$\lambda = \frac{340 \text{ m/s}}{60.0 \times 10^3 \text{ 1/s}}$$

$$\lambda = 5.67 \times 10^{-3} \text{ m} = 0.567 \text{ cm}$$

Following up

The simple derivation above shows that by using echo location a bat can detect insects roughly as small as half a centimeter across.

Using similar reasoning, what frequencies would you think would be useful for a dolphin in the ocean? The speed of sound in seawater is roughly 1530 m/s. What might be the typical minimum size of objects of interest to a dolphin? Without solving the equation explicitly, one can estimate that with the same frequency of sound as a bat, a dolphin in the ocean could detect an object about five times as big, or 2.5 cm (5 times the wave speed implies 5 times the size). A frequency of 20 kHz, which is toward the upper range of human hearing, would limit the dolphin to objects larger than 7.5 cm (1/3 the frequency implies 3 times the size).

PROBLEM 51 Frequency, Wavelength, and Wave Speed

Before we begin...

1. The speed of a wave v is related to the wavelength and frequency by

$$v = f\lambda$$

2. The known values are

$$f_{\text{low}} = 28 \text{ Hz} = 28 \text{ sec}^{-1}$$
$$f_{\text{high}} = 4200 \text{ Hz} = 4200 \text{ sec}^{-1}$$
$$v = 343 \text{ m/s}$$

Solving the problem

3. Solving the equation from step 1 for λ,

$$\lambda = \frac{v}{f}$$

4. Substituting the known values gives

$$\lambda_{\text{low-}f} = \frac{343 \text{ m/s}}{28 \text{ 1/s}} = 12 \text{ m}$$

$$\lambda_{\text{high-}f} = \frac{343 \text{ m/s}}{4200 \text{ 1/s}} = 8.2 \text{ cm}$$

The wavelengths range from approximately 8 cm to 12 meters.

PROBLEM 52 Mathematical Nature of Waves

Before we begin...

1. The given information is:

$A = 0.2$ m $\quad \lambda = 0.35$ m
$f = 12.0$ Hz $\quad \lambda(0, 0) = -0.03$ m

2.
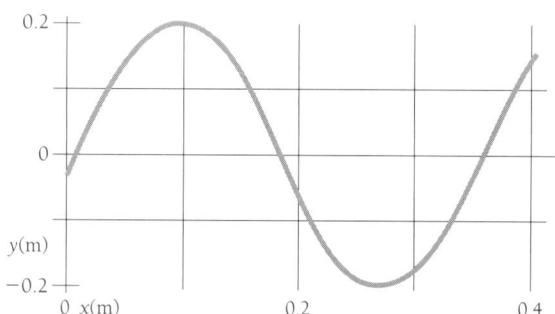

You must know the maximum and minimum values for y, the x distance before y repeats itself in phase, the position at a given point, and the sign of the velocity at that point, in order to create the plot.

3. The general expression for the wave traveling to the left as a function of position and time is given by

$$y(x, t) = A \cos(kx + \omega t + \phi)$$

4. The y position at $t = 0$ and $x = 0$ will be $A \cos \phi$. When the value of y at that point is known, this may be solved for ϕ.

Solving the problem

5. Using the values of wavelength λ and frequency f to find the wave number k and angular frequency ω, we have

$$k = 2\pi/\lambda = 18.0 \text{ rad/m}$$

$$\omega = 2\pi f = 75.4 \text{ rad/s}$$

We may also find the period T and phase velocity v

$$T = 1/f = 0.0833 \text{ s}$$

$$v = f\lambda = 4.20 \text{ m/s}$$

6. Substituting $t = 0$ and $x = 0$ into the general expression for the wave yields

$$y(0, 0) = A \cos(k(0) + \omega(0) + \phi)$$

Solving for ϕ

$$\phi = \cos^{-1} \frac{y(0, 0)}{A}$$

$$\phi = \cos^{-1} \frac{-0.03 \text{ m}}{0.2 \text{ m}} = \cos^{-1}(-0.15)$$

There are two solutions in the interval $-\pi \leq \phi \leq \pi$.

$$\phi = \pm 1.72 \text{ rad}$$

At the point (0, 0), the wave has a negative displacement and is traveling to the left. Since the wave has a positive velocity here (the displacement y is increasing with time), the correct solution for ϕ must satisfy

$$\cos(\phi + \text{"a little bit"}) > \cos(\phi)$$

For this to be true, $-\pi \leq \phi \leq 0$

or $\phi = -1.72$ rad

7. This particular traveling wave, using the general form and the computed values, is

$$y(x, t) = (0.2 \text{ m}) \cos[(18.0 \text{ rad/m})x + (75.4 \text{ rad/s})t - 1.72 \text{ rad}]$$

Core Concepts in College Physics Workbook

PROBLEM 53 Mathematical Nature of Waves

Before we begin...

1. The expression for the wave function of a traveling wave is

$$y(x, t) = A \cos(kx - \omega t + \phi)$$

Remember that $\cos(kx - \omega t - \pi/2) = \sin(kx - \omega t)$.

2. The wave described in the general case is traveling in the +x direction.

Solving the problem

3. A term-by-term comparison of the specific equation and the general equation yields

$A = 0.25$ m $k = 0.30$ m^{-1}

$\omega = 40$ s^{-1}

4. Using $k = 2\pi/\lambda$, we find

$\lambda = 21$ m

and with $\omega = 2\pi f$, the wave speed is

$$v = \left(\frac{\omega}{2\pi}\right)\left(\frac{2\pi}{k}\right) = \frac{\omega}{k} = 133 \text{ m/s}$$

5. An inspection of the sign in front of the time component of the wave function tells us that the wave is moving to the right (in the +x direction).

PROBLEM 54 Hooke's Law and the Equation of Motion

Before we begin...

1. The given variables and terms are

$m = 1.0$ kg $k = 25$ N/m $A = 3$ cm $= 0.03$ m

2. To solve the problem, we must (a) compute the period, (b) compute the maximum speed and maximum acceleration and (c) write equations expressing displacement, velocity, and acceleration as functions of time.

Solving the problem

3. We know that

$$\omega = \sqrt{\frac{k}{m}} = 5.0 \text{ rad/s} \quad \text{and} \quad T = \frac{2\pi}{\omega}$$

so $T = 1.26$ s.

4. Setting $\frac{1}{2}mv^2 = \frac{1}{2}kA^2$ and solving for v, we find

$$v = \sqrt{\frac{k}{m}} A$$

$$= 15 \text{ cm/s} = 0.15 \text{ m/s}$$

5. Setting $ma = -kx = kA$ and solving for a, we find that

$$a = 75 \text{ cm/s}^2 = 0.75 \text{ m/s}^2$$

6. Substituting the given value of A and the computed value of ω into the equation for position as a function of time, we have to evaluate the initial conditions to find ϕ. The spring will be at $x = -3.0$ cm $= -0.03$ m at $t = 0$, so $\phi = \pi$ radians.

$$x(t) = 0.03 \cos(5t + \pi) \text{ m}$$
$$v(t) = -0.15 \sin(5t + \pi) \text{ m/s}$$
$$a(t) = -0.75 \cos(5t + \pi) \text{ m/s}^2$$

All expressions and values have been computed, and the problem is solved. It is worth pointing out that had we solved for the general expressions before solving for the maximum values of velocity and acceleration, the answers for maximum values could have been found by inspection. The maximum value of a sine or cosine function is ± 1.0. Thus the maximum velocity is 0.15 m/s and the maximum acceleration is 0.75 m/s^2.

Module 8 **Simple Harmonic Motion**

PROBLEM 55 SHM and Waves in the Real World

Before we begin...

1. For small amplitudes, the relation between the period of motion of a simple pendulum, the pendulum's length, and the acceleration due to gravity is

$$T = 2\pi\sqrt{\frac{l}{g}}$$

2. The change in a quantity as measured under two different conditions is calculated by

$$\Delta T = T_2 - T_1$$

Solving the problem

3. Calculating the periods for each of the given acceleration values, we find

$$T_2 = 2\pi\sqrt{\frac{3.00 \text{ m}}{9.79 \text{ m/s}^2}} \quad T_1 = 2\pi\sqrt{\frac{3.00 \text{ m}}{9.80 \text{ m/s}^2}}$$

4. Subtracting the value computed for $g = 9.80$ m/s² from the value computed when $g = 9.79$ m/s²,

$$\Delta T = 2\pi\sqrt{l}\left(\frac{1}{\sqrt{g_2}} - \frac{1}{\sqrt{g_1}}\right)$$

$$= 1.78 \times 10^{-3} \text{ s}$$

PROBLEM 56 Wave Speed

Before we begin...

1. If the longitudinal wave travels the same or less distance, it would have to arrive first, since its speed is greater.

2. To calculate the time difference, the individual travel times t_t and t_l must be found. The speed v of each wave is known. To find the time t, we could use the equation

$$d = vt$$

where d is the distance traveled. To use this equation, the distance from A to B over each path must be found.

3.

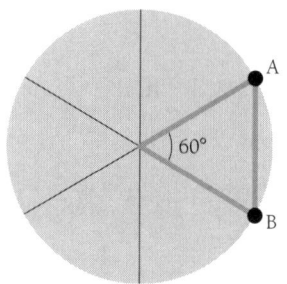

4. The given values are

$$R = 6.37 \times 10^6 \text{ m}$$
$$v_t = 4.50 \text{ km/s}$$
$$v_l = 7.80 \text{ km/s}$$

Solving the problem

5. The arc $\widehat{AB}$ covers 60° of the circumference C of the Earth.

$$\widehat{AB} = \frac{60°}{360°}C = \frac{1}{6}(2\pi R)$$

The direct line from A to B is one leg of an equilateral triangle, with each of the other legs being a radius of the earth. Its length is

$$\overline{AB} = 2R \sin(\theta/2) = R$$

6. Solving the equation in step 2 for t, and substituting for the distance gives, for the transverse wave,

$$t_t = \frac{\widehat{AB}}{v_t} = \frac{(1/6)(2\pi R)}{v_t}$$

and, for the longitudinal wave, the same steps yield

$$t_l = \frac{\overline{AB}}{v_l} = \frac{R}{v_l}$$

7. Substituting the known values gives

$$t_t = \frac{(1/6)(2\pi)(6.37 \times 10^6 \text{ m})}{4.50 \text{ km/s}}$$
$$= 1.48 \times 10^3 \text{ sec}$$

and

$$t_l = \frac{(6.37 \times 10^6 \text{ m})}{7.80 \text{ km/s}}$$
$$= 8.17 \times 10^2 \text{ sec}$$

The longitudinal wave takes less time. The difference in the arrival times is

$$\Delta t = 660 \text{ sec} = 11 \text{ minutes}$$

Following up

If A and B were only 200 kilometers apart on the surface, then the travel time for the tranverse wave would be

$$t_t = \frac{200 \text{ km}}{4.50 \text{ km/s}} = 44 \text{ sec}$$

They would be 1.8° apart in latitude

$$\frac{(360°/2\pi)(2.0 \times 10^5 \text{ m})}{6.37 \times 10^6 \text{ m}}$$

The direct path through the Earth would be

$$\overline{AB} = 2R \sin(0.9°) = 204 \text{ km}$$

$$t_l = \frac{204 \text{ km}}{7.80 \text{ km/s}} = 26 \text{ sec}$$

The time difference is then 18 seconds. For distances close enough to feel an earthquake, the difference in travel times of the two kinds of waves can be roughly estimated to be 0.1 second for every kilometer away from the epicenter. If one feels the two kinds of waves, one can roughly calculate the distance away from the epicenter by estimating the time between the arrivals of each kind of shaking.

Module 9 Wave Behavior

PROBLEM 57 Speed of a Wave in a Medium

Before we begin...

1. The given information is

$$\mu = 8.00 \text{ g/m} = 0.00800 \text{ kg/m}$$

$$v = 60.0 \text{ m/s}$$

2.

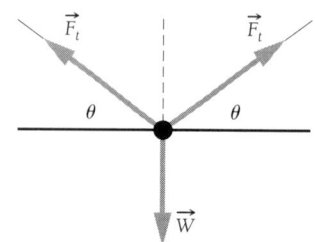

Solving the problem

3. The tension in each section of the string is the same, ensuring that the x components of the forces equal zero, because the tensions make the same angle with respect to the x axes. The two y components together will be equal to the weight.

4. From the original drawing of the problem, we use one of the right triangles formed by the string and the span of the supports to get

$$\theta = \cos^{-1}\left(\frac{3L/8}{L/2}\right) = 41.4°$$

5. Applying the condition of equilibrium to the y components and solving for the tension F_t

$$F_t \sin \theta + F_t \sin \theta = mg$$

$$2F_t \sin \theta = mg$$

$$F_t = \frac{mg}{2 \sin \theta}$$

6. Substituting the above expression for F_t into the relation, we obtain

$$v = \sqrt{\frac{mg}{2\mu \sin \theta}}$$

7. Solving for m,

$$m = \frac{2\mu v^2 \sin \theta}{g}$$

Substituting known values,

$$m = \frac{2(0.008 \text{ kg/m})(60.0 \text{ m/s})^2 \sin 41.4°}{9.8 \text{ m/s}^2}$$

$$= 3.9 \text{ kg}$$

Core Concepts in College Physics Workbook

PROBLEM 58 Energy and Power in Waves

Before we begin…

1. The power delivered by a transverse wave is expressed by

$$P = \frac{1}{2}\mu\omega^2 A^2 v$$

2. The length of the rope does not affect the power.

Solving the problem

3. If the speed remains constant and only the length changes, the power does not change.

If A is doubled and ω is halved, the power does not change, because $P \propto \omega^2 A^2$.

Because $\omega \propto 1/\lambda$, doubling λ will cut ω in half, just as in the previous question. Thus, the power does not change when both λ and A are doubled.

If the wavelength λ is halved, ω will double, so ω^2 will increase fourfold. The power will quadruple.

PROBLEM 59 Superposition and Interference

Before we begin…

1. The wave function of each wave is

$$y_1 = (4.0 \text{ cm}) \sin(kx - \omega t)$$

$$y_2 = (4.0 \text{ cm}) \sin(kx - \omega t - \phi)$$

2. The superposition principle states that the resultant wave function of two or more traveling waves in a medium is the sum of the wave functions of the individual waves.

$$y = y_1 + y_2$$

Solving the problem

3. Substituting into the superposition principle and using the trig identity relating $\sin(a) + \sin(b)$, we find

$$y = (4.0 \text{ cm}) \sin(kx - \omega t) \\ + (4.0 \text{ cm}) \sin(kx - \omega t - 90°) \\ = 2(4.0 \text{ cm}) \sin(kx - \omega t - 45°) \cos 45°$$

If we rearrange the equation, we can write it as

$$y = [(2)(4.0 \text{ cm}) \cos 45°] \sin(kx - \omega t - 45°)$$

Inspection of the form of the equation tells us that the amplitude is

$$A = [(2)(4.0 \text{ cm}) \cos 45°] = 5.7 \text{ cm}$$

PROBLEM 60 Standing Waves

Before we begin...

1. Node positions are stationary points on the resulting wave; that is, places where $y(x, t) = 0$ for all values of t.

2. The sum of the displacements of all the waves must result in a zero displacement in order for a point to be a node.

Solving the problem

3. Using the superposition principle to find the wave function of the combined waves, we find

$$y(x, t) = y_1(x, t) + y_2(x, t)$$
$$= A \sin(kx - \omega t) + A \sin(2kx + \omega t)$$
$$= 2A \cos\left(\frac{2kx + \omega t - kx + \omega t}{2}\right)$$
$$\times \sin\left(\frac{2kx + \omega t + kx - \omega t}{2}\right)$$

The above equation simplifies to

$$y(x, t) = 2A \cos\left(\frac{kx + 2\omega t}{2}\right) \sin\left(\frac{3kx}{2}\right)$$

4. Stationary nodes occur as a result of a term that is independent of time t equaling zero. This is true when

$$\sin\left(\frac{3kx}{2}\right) = 0$$

For this to be satisfied, $3kx = 0, \pm 2\pi, \pm 4\pi \cdots$, therefore $x = 0, 2\pi/3k, 4\pi/3k$, etc.

5. Even at points which are not nodes, $y(x, t)$ will be zero whenever

$$\cos\left(\frac{kx + 2\omega t}{2}\right) = 0$$

The solutions to this equation are

$$x = \pm(\pi - 2\omega t)/k,$$
$$\pm(3\pi - 2\omega t)/k,$$
$$\pm(5\pi - 2\omega t)/k, \text{ etc.}$$

PROBLEM 61 Standing Waves—Wave Fixed at Both Ends

Before we begin...

1. The given information is

$$M = 0.10 \text{ kg} \qquad L = 2.0 \text{ m}$$
$$F_t = 20 \text{ N}$$

2. The equation for the allowable wavelengths for a wave fixed at both ends is

$$\lambda_n = 2L/n \quad (n = 1, 2, 3,...)$$

3. The speed of the wave can be calculated for a wire under tension using

$$v = \sqrt{\frac{F_t}{\mu}}$$

Solving the problem

4. The mass density is

$$\mu = \frac{M}{L} = \frac{0.10 \text{ kg}}{2.0 \text{ m}} = 0.050 \text{ kg/m}$$

Using the value of μ and the given tension,

$$v = \sqrt{\frac{20 \text{ N}}{0.050 \text{ kg/m}}} = 20 \text{ m/s}$$

5. Substituting λ_n in the relationship $f\lambda = v$ and solving for f,

$$f_n = \frac{v}{\lambda_n}$$
$$= \frac{nv}{2L}$$

6. For the first mode, $n = 1$,

$$f_1 = \frac{20 \text{ m/s}}{4 \text{ m}} = 5.0 \text{ Hz}$$

Other nodes have

$$f_n = nf_1$$

so $f_2 = 10.0$ Hz, $f_3 = 15.0$ Hz, ...

Core Concepts in College Physics Workbook

PROBLEM 62 Standing Waves—Wave with One Fixed End and One Free End

Before we begin...

1. The relationship between the length of the system and the allowable frequencies for this type of system is

$$f_n = n \frac{v}{4L} \quad (n = 1, 3, 5, ...)$$

This can be rewritten in terms of n representing the mode number of the vibration as

$$f_n = (2n - 1) \frac{v}{4L} \quad (n = 1, 2, 3, ...)$$

2. The given information is

$$f_a = 52.0 \text{ Hz} \qquad f_b = 60.0 \text{ Hz}$$

Solving the problem

3. Writing the expressions for the nth allowable frequency for $a = n$ and $b =$ next allowable frequency, we find

$$f_a = (2n - 1)\frac{v}{4L} \qquad f_b = [2(n+1) - 1]\frac{v}{4L}$$

4. The simultaneous equations to solve are written as

$$f_b - f_a = 60.0 \text{ Hz} - 52.0 \text{ Hz} = 8.0 \text{ Hz}$$

and

$$f_b - f_a = [2(n+1) - 1]\frac{v}{4L} - (2n - 1)\frac{v}{4L}$$

5. After simplifying, the solution gives us the expression

$$8.0 \text{ Hz} = \frac{2v}{4L}$$

from which we find $L = 21.5$ m, the distance to the water surface.

Module 10 Thermodynamics

PROBLEM 63 Basic Concepts of Thermodynamics

Before we begin...

1. Thermal equilibrium is reached when no net heat flows from one object to the other. Objects in thermal equilibrium are said to be at the same temperature.

2. The given information is

$$V_w = 500 \text{ ml} \qquad T_{1w} = 30° \text{ C}$$
$$m_{ice} = 25 \text{ g} \qquad T_{1\text{-ice}} = 0° \text{ C}$$

Solving the problem

3. The mass of the water is

$$m = (10^3 \text{ kg/m}^3)(1 \text{ m}^3/10^3 \text{ l})(0.500 \text{ l})$$
$$= 0.500 \text{ kg}$$

4. The thermal energy that would be released by lowering the water to the freezing point is

$$Q = mc\Delta T = (0.500 \text{ kg})(4186 \text{ J/kg C°})(0° \text{ C} - 30° \text{ C})$$
$$= -6.28 \times 10^4 \text{ J}$$

(The minus sign indicates that energy is released rather than absorbed.)

5. The thermal energy required to melt all of the ice is

$$Q = mL_f = (0.0250 \text{ kg})(3.33 \times 10^5 \text{ J/kg})$$
$$= 8.33 \times 10^3 \text{ J}$$

6. Because the water has more available thermal energy than that required to melt the ice, the water does not cool all the way to 0° C. Once all the ice has melted, the combined pool of water will reach equilibrium at some intermediate temperature T_2, which we can find using energy conservation

$$mc(T_2 - 30°C) + 8.33 \times 10^3 \text{ J}$$
$$+ m_{ice}c(T_2 - 0° \text{ C}) = 0$$

$$T_2 = 24.6° \text{ C}$$

PROBLEM 64 The Ideal Gas

Before we begin...

1. The ideal gas law is expressed by the relationship

$$PV = Nk_B T$$

where P is the pressure in atmospheres, V the volume, N the number of gas molecules present, and T the temperature in Kelvins. The term k_B is a constant called Boltzmann's constant.

2. The given information is

$$T_1 = 10.0° \text{ C} \quad T_2 = 80.0° \text{ C} \quad P_1 = 2.50 \text{ atm}$$

Solving the problem

3. On the Kelvin scale, the temperatures T_1 and T_2 are

$$T_1 = 10.0° + 273° = 283 \text{ K}$$

and

$$T_2 = 80.0° + 273° = 353 \text{ K}$$

4. Collecting the relevant quantities P and T on one side of the ideal gas law,

$$\frac{P}{T} = \frac{Nk_B}{V}$$

Because V and N are constant in this problem, we can write

$$\frac{P_1}{T_1} = \frac{P_2}{T_2}$$

Solving for P_2 we find

$$P_2 = \frac{P_1 T_2}{T_1} = \frac{(2.50 \text{ atm})(353 \text{ K})}{(283 \text{ K})}$$

$$= 3.12 \text{ atm}$$

PROBLEM 65 The First Law of Thermodynamics

Before we begin...

1. This first law of thermodynamics for a gas can be expressed

$$\Delta U = U_f - U_i = Q - W$$

2. In general, during an expansion, a gas does work equal to the area under the curve which traces its pressure and volume values from its initial state to its final state.

3. At constant pressure, the work done is

$$W = P\Delta V$$

4. Work is force exerted over some distance. If the volume is not changing, then no work is done, no matter what the pressure change.

5. The given information is

initial pressure of the gas $P_i = 2.0$ atm
initial volume of the gas $V_i = 0.30$ L
initial internal energy $U_i = 91$ J
final pressure of the gas $P_f = 1.5$ atm
final volume of the gas $V_f = 0.80$ L
final internal energy $U_f = 182$ J

Solving the problem

6. Breaking the path IAF into its isobaric and isovolumetric parts gives an expression for the work done by the gas

$$W_{IAF} = W_{IA} + W_{AF} = 0 + P_f(V_f - V_i)$$

The corresponding expression for IBF is

$$W_{IBF} = W_{IB} + W_{BF} = P_i(V_f - V_i) + 0$$

Core Concepts in College Physics Workbook

7. Since the work done is the area under the curve on the PV diagram, the work done on the path IF equals the work done on path IAF plus the area of the triangle IAF. But that area is exactly half of the square IBFA, the difference between the work done on path IBF and the work done on path IAF. Therefore,

$$W_{IF} = \frac{W_{IAF} + W_{IBF}}{2}$$

8. For every path,

$$Q = (U_f - U_i) + W$$

9. Substituting the known values results in the following values for the work along each path:

$$W_{IAF} = (1.5 \text{ atm})(0.80 - 0.30)\text{L} = 0.75 \text{ atm·L}$$

$$W_{IBF} = (2.0 \text{ atm})(0.50 \text{ L}) = 1.0 \text{ atm·L}$$

$$W_{IF} = \frac{W_{IAF} + W_{IBF}}{2} = 0.88 \text{ atm·L}$$

Converting units,

$$1 \text{ atm} = 1.01 \times 10^5 \text{ Pa},\ 1 \text{ L} = 10^{-3} \text{ m}^3$$

$$W_{IAF} = 0.75 \text{ atm·L} = 76 \text{ Pa·m}^3 = 76 \text{ J}$$

$$W_{IBF} = 1.0 \text{ atm·L} = 101 \text{ J}$$

$$W_{IF} = \frac{W_{IAF} + W_{IBF}}{2} = 89 \text{ J}$$

Substituting our expressions for work and the values for internal energy yields the heat absorbed along each path:

$$Q_{IAF} = (182 \text{ J} - 91 \text{ J}) + 76 \text{ J} = 167 \text{ J}$$

$$Q_{IBF} = (182 \text{ J} - 91 \text{ J}) + 101 \text{ J} = 192 \text{ J}$$

$$Q_{IF} = (182 \text{ J} - 91 \text{ J}) + 89 \text{ J} = 180 \text{ J}$$

PROBLEM 66 Carnot Engines

Before we begin...

1. The given information is

$$T_c = 20°\text{ C} \qquad T_h = 500°\text{ C} \qquad P = 150 \text{ kW}$$

2. Power P is the rate at which work is done.

$$P = \frac{\Delta W}{\Delta t}$$

3. The efficiency of any heat engine is expressed by the relationship

$$e = \frac{W}{Q_h}$$

where W is the work done by the engine and Q_h is the heat absorbed from its hot reservoir.

4. For a Carnot engine, the efficiency is given by

$$e_C = \frac{(T_h - T_c)}{T_h}$$

where all temperatures are in kelvins.

Solving the problem

5. Solving for the efficiency of the engine we find

$$e_C = \frac{(T_h - T_c)}{T_h} = \frac{480 \text{ K}}{773 \text{ K}} = 0.621$$

6. The rate at which work is being done by the engine is

$$1.50 \times 10^5 \text{ W}$$

so in one hour (3600 s) the work will be

$$W = (1.50 \times 10^5 \text{ W})(3600 \text{ s}) = 5.40 \times 10^8 \text{ J}$$

7. The heat absorbed will be

$$Q_h = W/e_C = 8.70 \times 10^8 \text{ J}$$

8. Because $W = Q_h - Q_c$, we compute

$$Q_c = 8.70 \times 10^8 \text{ J} - 5.40 \times 10^8 \text{ J}$$

$$= 3.30 \times 10^8 \text{ J}$$

PROBLEM 67 Carnot Engines—The Heat Pump

Before we begin...

1. The Carnot COP is computed by the relationship

$$\text{COP} = \frac{Q_h}{W} = \frac{Q_h}{Q_h - Q_c} = \frac{T_h}{T_h - T_c}$$

2. The given information is

$$T_h = 22°\text{ C} \qquad T_c = -3°\text{ C}$$

Solving the problem

3. The temperatures T_h and T_c on the Kelvin scale are

$$T_h = 22°\text{ C} = 295\text{ K} \quad \text{and} \quad T_c = -3°\text{ C} = 270\text{ K}$$

4. We can evaluate the COP as

$$\frac{T_h}{T_h - T_c} = \frac{295\text{ K}}{25\text{ K}} = 11.8$$

PROBLEM 68 Entropy

Before we begin...

1. When an amount of heat ΔQ flows into a system at temperature T (measured on an absolute scale such as Kelvin's), the change in the system's entropy is

$$\Delta S = \frac{\Delta Q}{T}$$

2. Every phase change involves a change in internal energy. The heat required as a liquid of mass m changes to a gas is

$$Q_v = mL_v$$

where L_v is the latent heat of vaporization.

The heat required as a mass m changes from a solid to a liquid is

$$Q_f = mL_f$$

L_f is the latent heat of fusion.

Heat must be absorbed by the water as it changes from liquid to gas, as the gaseous molecules have more kinetic energy.

To freeze water to ice, heat must be removed from the water by the freezer.

3. The given information is:

$$m_a = 1\text{ kg}$$
$$T_a = 100°\text{ C}$$
$$V_b = 1\text{ L}$$
$$T_b = 0°\text{ C}$$

Solving the problem

4. Converting from Celsius to Kelvin scales,

$T_a = (100° + 273°)\ K = 373°\ K$

$T_b = (0° + 273°)\ K = 273°\ K$

5. The mass in kilograms of the water frozen is

$m_b = \rho V_b = (1\text{g/cm}^3)(1\ \text{L})$

$= \left(\dfrac{1\ \text{g}}{\text{cm}^3}\right)(1\ \text{L})\left(\dfrac{1000\ \text{cm}^3}{\text{L}}\right)\left(\dfrac{1\ \text{kg}}{1000\text{g}}\right)$

$m_b = 1\ \text{kg}$

6. Substituting the first equation from step 2 into the equation from step 1 yields the change of entropy when the water changes to steam in terms of the mass, the heat of vaporization, and the temperature.

$\Delta S_a = \dfrac{\Delta Q_a}{T} = \dfrac{Q_v}{T} = \dfrac{mL_v}{T}$

$= \dfrac{(1\ \text{kg})(2.26 \times 10^6\ \text{J/kg})}{373°\text{K}}$

$= 6.06 \times 10^3\ (\text{J/°K})$

Using the second equation from step 2, the heat for a solid to liquid phase change (remembering that in freezing the transition is liquid to solid) yields the entropy change as the water freezes, in terms of the mass, the heat of fusion, and the temperature.

$\Delta S_b = \dfrac{\Delta Q_b}{T} = \dfrac{-Q_f}{T} = \dfrac{-mL_f}{T}$

$= \dfrac{-(1\ \text{kg})(3.33 \times 10^5\ \text{J/kg})}{273°\text{K}}$

$= 1.22 \times 10^3\ (\text{J/°K})$

7. By conservation of energy, the change in energy of the system of the freezer and water must be zero (i.e., the heat absorbed by the freezer must equal the heat flowing out of the water as it freezes):

$\Delta Q_\text{freezer} + \Delta Q_b = 0$:

The change in entropy of the freezer is therefore

$\Delta S_\text{freezer} = \dfrac{\Delta Q_\text{freezer}}{T} = \dfrac{-\Delta Q_b}{T} = -\Delta S_b$

$= 1.22 \times 10^3\ (\text{J/°K})$

Following up

The second law of thermodynamics is sometimes stated as "the entropy of the Universe increases in all natural processes." The second law might seem to be violated as water freezes since the change in entropy of the water exactly equals the change in entropy of the freezer. We have assumed that the temperature of the freezer is the same as the water. This is a simplification. No net heat transfer can take place between them if they are at the same temperature. If the temperature of the freezer is less than the water, then the entropy change of the freezer will be larger in magnitude than that of the water, and the entropy of the Universe will increase as expected. If some other process (such as the device in a real freezer) is extracting heat from the freezer at the same rate heat is being transferred from the water, than that process will contribute a net increase in entropy.

Module 10 **Thermodynamics**

Module 11 The Electric Field

PROBLEM 69 Coulomb's Law

Before we begin...

1. The given information is

$q = +10.0\ \mu C$ $L = 60.0$ cm $= 0.600$ m

$W = 15.0$ cm $= 0.150$ m

2. Coulomb's law for the force between two point charges states that the two charges will experience a force that is described by the relationship

$$F_{12} = k_e \frac{q_1 q_2}{r^2}$$

where k_e is a constant and r is the distance between the charges. The force is along the line joining the two charges.

3. The direction of the force is attractive when the charges are of opposite sign. If the charges are of like sign, the force is repulsive.

4.

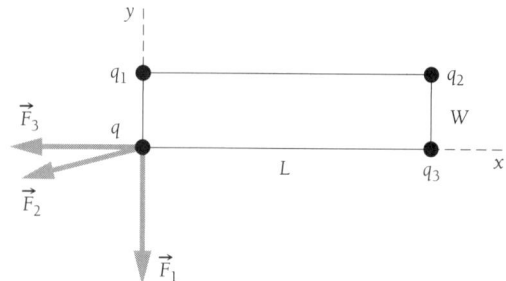

Solving the problem

5. Using Coulomb's law, we can evaluate the forces $\vec{F}_1$, $\vec{F}_2$, and $\vec{F}_3$ as

$$F_1 = k_e \frac{q_1 q}{r_1^2}$$
$$= (9.00 \times 10^9\ \text{N·m}^2/\text{C}^2)$$
$$\times \frac{(1.00 \times 10^{-5}\ \text{C})(1.00 \times 10^{-5}\ \text{C})}{(0.150\ \text{m})^2}$$
$$= 40.0\ \text{N}$$

The direction of $\vec{F}_1$ is 270° with respect to the x direction because the charges repel each other.

$$F_2 = \frac{k_e q_2 q}{r_2^2}$$
$$= (9.00 \times 10^9\ \text{N·m}^2/\text{C}^2)$$
$$\times \frac{(1.00 \times 10^{-5}\ \text{C})(1.00 \times 10^{-5}\ \text{C})}{(0.600\ \text{m})^2 + (0.150\ \text{m})^2}$$
$$= 2.35\ \text{N}$$

To find the direction, we evaluate

$$\theta = \tan^{-1} \frac{0.150\ \text{m}}{0.600\ \text{m}} = 14°$$

The direction of $\vec{F}_2$ is $180° + 14° = 194°$ with respect to the $+x$ axis because the charges repel each other.

$$F_3 = k_e \frac{q_3 q}{r_3^2}$$
$$= (9.00 \times 10^9\ \text{N·m}^2/\text{C}^2)$$
$$\times \frac{(1.00 \times 10^{-5}\ \text{C})(1.00 \times 10^{-5}\ \text{C})}{(0.600\ \text{m})^2}$$
$$= 2.50\ \text{N}$$

The direction of $\vec{F}_3$ is 180°.

6. Resolving the forces into their x and y components, we find

$F_{1x} = 0$ $F_{1y} = -40.0$ N

$F_{2x} = -2.28$ N $F_{2y} = -0.569$ N

$F_{3x} = -2.50$ N $F_{3y} = 0$

so that

$F_{Tx} = -4.78$ N $F_{Ty} = -40.6$ N

7. The magnitude of $\vec{F}_T$ is

$$F_T = \sqrt{F_{Tx}^2 + F_{Ty}^2}$$
$$= \sqrt{(-4.78)^2 + (-40.6)^2} = 40.9\ \text{N}$$

8. The direction of $\vec{F}_T$ is

$$\theta = \tan^{-1} \frac{F_y}{F_x} = \tan^{-1} \frac{-40.7}{-4.78} = 263°$$

PROBLEM 70 The Electric Field and Field Lines

Before we begin...

1.

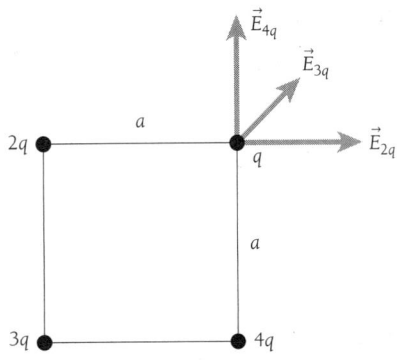

2. The electric field for a point charge is calculated using the relationship

$$\vec{E} = \frac{\vec{F}_e}{q_o}$$

Using Coulomb's law for the force,

$$|\vec{E}| = k_e \frac{q}{r^2}$$

3. The direction of the electric field due to a point charge points radially away from a positive charge creating the field and radially toward a negative charge creating the field.

4. The electric force on a charge q is given by $\vec{F} = q\vec{E}$.

Solving the problem

5. Evaluating the fields $\vec{E}_{2q}$, $\vec{E}_{3q}$ and $\vec{E}_{4q}$, we find

$$E_{2q} = k_e \frac{(2q)}{a^2}$$

(The direction of $\vec{E}_{2q}$ is 0° or along the +x axis because the charge is positive, thus the field points away from the charge.)

$$E_{3q} = k_e \frac{(3q)}{(\sqrt{2}a)^2}$$

(The direction of $\vec{E}_{3q}$ is 45° because the charge is positive.)

$$E_{4q} = k_e \frac{(4q)}{a^2}$$

(The direction of $\vec{E}_{4q}$ is 90° because the charge is positive.)

6. Resolving the fields into their x and y components we find

$$E_{2qx} = k_e \frac{(2q)}{a^2} \qquad E_{2qy} = 0$$

$$E_{3qx} = k_e \frac{(3q)(\cos 45°)}{(\sqrt{2}a)^2}$$

$$E_{3qy} = k_e \frac{(3q)(\sin 45°)}{(\sqrt{2}a)^2}$$

$$E_{4qx} = 0 \qquad E_{4qy} = k_e \frac{(4q)}{a^2}$$

therefore

$$E_{Tx} = \frac{(3.06\, k_e)q}{a^2} \quad \text{and}$$

$$E_{Ty} = \frac{(5.06\, k_e)q}{a^2}$$

7. Calculating the magnitude of $\vec{E}_T$, we find

$$E_T = \sqrt{E_{Tx}^2 + E_{Ty}^2}$$
$$= \frac{k_e q}{a^2}\sqrt{(3.06)^2 + (5.06)^2} = 5.91 \frac{k_e q}{a^2}$$

8. The direction is determined to be

$$\theta = \tan^{-1}\frac{E_y}{E_x} = 58.8°$$

9. The electric force $\vec{F}$ points in the same direction as $\vec{E}_T$ (if q is positive) and has magnitude

$$F = qE_T = 5.91 \frac{k_e q^2}{a^2}$$

PROBLEM 71 Gauss's Law

Before we begin...

1. Gauss's law for electric fields states that the total electric flux is equal to the net enclosed charge divided by a constant.

$$\Phi = \frac{q_{in}}{\epsilon_0}$$

2. Because the electric flux does not depend upon the shape of the closed surface, the charge can be calculated from the flux without regard to the type of surface. It is only when we are evaluating the electric field $\vec{E}$ that the type of surface is selected for symmetry.

3. If the electric flux is negative, $\vec{E}$ and $\vec{n}$ must point in opposite directions. Under this condition, the net charge within the surface must be negative.

Solving the problem

4. Using the relationship between total electric flux and the net charge, we find

$$q_{in} = \epsilon_0 \Phi = (8.85 \times 10^{-12} \text{ C}^2/\text{N·m}^2)$$
$$\times (8.60 \times 10^4 \text{ N·m}^2/\text{C})$$
$$= 7.61 \times 10^{-7} \text{ C}$$

Because the flux is positive, the net charge within the surface must also be positive.

If the flux had the same magnitude but were negative, the magnitude of the charge enclosed by the surface would not change. The sign of the charge, however, would be negative.

PROBLEM 72 Examples of the Electric Field

Before we begin...

1. The charge inside a right cylinder of length l and radius r would be

$$q_{in} = \lambda l$$

2. The flux through the two end caps of the cylinder will equal zero because $\vec{E}$ and $\vec{n}$ are perpendicular.

Solving the problem

3. Applying Gauss's law to derive the electric field at radial distances from a long straight wire, we have three surfaces over which to integrate: the two end caps and the side surface surrounding the filament. The end caps yield zero flux.

Evaluating the flux through the side surface, we have

$$\Phi = EA = E(2\pi rl)$$

because $\vec{E}$ on the cylindrical surface is constant and parallel to $\vec{n}$.

By setting the flux $\Phi = q_{in}/\epsilon_0$, substituting for Φ from above, and using $q_{in} = \lambda l$, we find

$$E(2\pi rl) = \frac{\lambda l}{\epsilon_0}$$

$$E = \frac{\lambda}{2\pi \epsilon_0 r}$$

4. In this problem, λ is negative, indicating the direction of $\vec{E}$ is toward the filament rather than away from it. Evaluating for $r_1 = 10.0$ cm $= 0.1$ m, we find

$$E_1 = \frac{(-90.0 \times 10^{-6} \text{C/m})}{(2\pi)(8.85 \times 10^{-12} \text{ C}^2/\text{N·m}^2)(0.100 \text{ m})}$$

$$= -1.62 \times 10^7 \text{ N/C}$$

Solving for E_2 and E_3, we find

$$E_2 = -8.10 \times 10^6 \text{ N/C}$$
$$E_3 = -1.62 \times 10^6 \text{ N/C}$$

PROBLEM 73 Electric Potential

Before we begin...

1. The given information is

   $q_1 = 500$ nC $= 5.00 \times 10^{-9}$ C
   $q_2 = -3.00$ nC $= -3.00 \times 10^{-9}$ C
   distance between charges $= 35$ cm $= 0.35$ m
   distance between A and $q_1 = 0.175$ m
   distance between A and $q_2 = 0.175$ m

2. Electric potential is given by

   $$V = k_e \frac{q}{r}$$

   To calculate the electric potential at a point due to more than one charge, simply add the individual potentials contributed by each charge.

3. The electric field at a point has magnitude

   $$E = k \frac{|q|}{r^2}$$

   and points directly away from a positive charge, or toward a negative charge. To calculate the electric field at a point due to more than one charge, simply add (as vectors) the individual electric fields at that point.

Solving the problem

4. The potential at point A due to each of the individual charges is

   $$V_1 = k_e \frac{q_1}{r}$$

   $$= \frac{(8.99 \times 10^9 \text{ N·m}^2/\text{C}^2)(5.00 \times 10^9 \text{ C})}{(0.175 \text{ m})}$$

   $$= 257 \text{ V}$$

   $$V_2 = k_e \frac{q_2}{r}$$

   $$= \frac{(8.99 \times 10^9 \text{ N·m}^2/\text{C}^2)(-3.00 \times 10^9 \text{ C})}{(0.175 \text{ m})}$$

   $$= -154 \text{ V}$$

   The total electric potential at point A is

   $$V_T = V_1 + V_2 = 103 \text{ V}$$

5. The total electric field $\vec{E}_T$ at a point equals the vector sum $\vec{E}_1 + \vec{E}_2$, where $\vec{E}_1$ is the field due to q_1 and $\vec{E}_2$ is the field due to q_2.

   $$\vec{E}_1 = k_e \frac{|q_1|}{r_1^2}$$

   $$= (8.99 \times 10^9 \text{ N·m}^2/\text{C}^2) \frac{|(5.00 \times 10^9 \text{ C})|}{(0.175 \text{ m})^2}$$

   $$= 1468 \text{ N/C, directed to the right (away from the positive charge } q_1)$$

   $$\vec{E}_2 = k_e \frac{|q_2|}{r_2^2}$$

   $$= (8.99 \times 10^9 \text{ N·m}^2/\text{C}^2) \frac{|(-3.00 \times 10^9 \text{ C})|}{(0.175 \text{ m})^2}$$

   $$= 881 \text{ N/C, directed to the right (toward the negative charge } q_2)$$

   Because $\vec{E}_1$ and $\vec{E}_2$ point in the same direction, we can find their vector sum by simply adding their magnitudes. $\vec{E}_T = \vec{E}_1 + \vec{E}_2 = 2349$ N/C, directed toward the right.

PROBLEM 74 Electric Field and Electric Potential

Before we begin...

1. When the separation between two parallel plates is small relative to the dimensions of the plates, the electric field will be very uniform between the plates.

2. In a uniform electric field, the component of the average electric field directed along the line between any two points is related to the electric potential difference by

$$E_{\|} = \frac{-\Delta V}{d}$$

where d is the distance between the points and $E_{\|}$ is the component of the electric field parallel to the line joining the points.

3. The relationship between force and field comes from the definition of the electric field

$$\vec{E} = \frac{\vec{F}}{q_0}$$

where q_0 is a small positive test charge.

When strength and direction of the field is known, the force on a particle can be calculated by rearranging this

$$\vec{F} = q\vec{E}$$

where now q is the particle's charge.

4.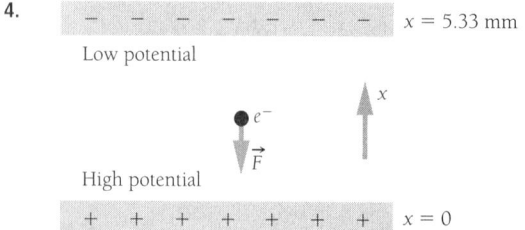

We have chosen the x axis perpendicular to the plates, with $x = 0$ on the positive plate, and $+x$ in the direction of the negative plate.

5. In general, work is defined as force acting over distance.

$$W = Fd$$

For the special case of a uniform electric field, when a particle moves a distance d parallel to the direction of the field, the work is given by

$$W = qEd$$

6. The known quantities are

$$\Delta V = -600 \text{ V}$$
$$s = 5.33 \times 10^{-3} \text{ m}$$
$$q = -1.6 \times 10^{-19} \text{ C}$$
$$d = 5.33 \times 10^{-3} \text{ m} - 2.90 \times 10^{-3} \text{ m}$$
$$= 2.43 \times 10^{-3} \text{ m}$$

(The electron starts out 2.90 mm from the positive plate, so it is 2.43 mm from the negative plate.)

Solving the problem

7. Substituting the known values into the equation from part 2, remembering that the distance here is the separation between plates, and that the potential decreases from $x = 0$ to $x = s$,

$$E = -\frac{-600 \text{ V}}{5.33 \times 10^{-3} \text{ m}}$$
$$= +1.13 \times 10^5 \text{ V/m}$$

The plus sign means the electric field is in the direction of the $+x$ axis (pointing from the positive plate to the negative plate).

8. Using the value for the electric field and the charge of the electron yields

$$F = (1.13 \times 10^5 \text{ V/m})(-1.6 \times 10^{-19} \text{ C})$$

Since 1 V/m = 1 N/C, in newtons the force is

$$F = -1.81 \times 10^{-14} \text{ N}$$

The force is directed toward the positive plate.

9. Using the definition of work from above, with the magnitude of the force and the distance the electron is moved

$$W_E = (-1.81 \times 10^{-14} \text{ N})(2.43 \times 10^{-3} \text{ m})$$
$$= -4.40 \times 10^{-17} \text{ N·m}$$
$$= -4.40 \times 10^{-17} \text{ J}$$

This is the work done by the field, but to move the electron a force must be applied which opposes the force from the field. Therefore the work done on the electron is

$$W = -W_E = 4.40 \times 10^{-17} \text{ J}$$

which is, of course, positive, since work must be done on the electron to move it against the force of the electric field.

Module 12 The Magnetic Field

PROBLEM 75 Magnetic Force on a Moving Charge

Before we begin...

1. The magnetic force $\vec{F}_B$ is related to the magnetic field $\vec{B}$ and the velocity $\vec{v}$ of a moving charged particle by the cross product equation

$$F_B = qvB \sin \theta$$

2. This force does not change the speed of the particle because it acts in a perpendicular direction to the velocity; therefore, it causes a centripetal acceleration.

3.

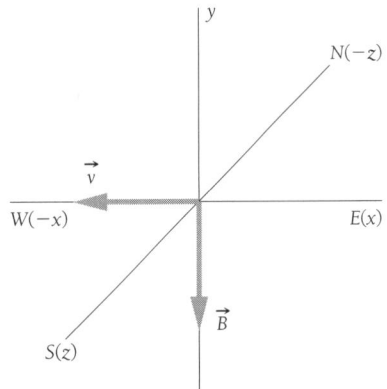

4. The given information is

$B = 0.5 \times 10^{-4}$ T $\qquad v = 6.2 \times 10^6$ m/s

$q = +1.60 \times 10^{-19}$ C $\qquad m = 1.67 \times 10^{-27}$ kg

(continued on next page . . .)

Solving the problem

5. $F_B = qvB \sin \theta$, where θ is the angle between $\vec{v}$ and $\vec{B}$. Using the right-hand rule, the direction of the initial force corresponds to southward.

$$F_B = [(+1.60 \times 10^{-19} \text{ C})$$
$$(6.2 \times 10^6 \text{ m/s})(0.5 \times 10^{-4} \text{ T}) \sin 90°]$$
$$= 5.0 \times 10^{-17} \text{ N}$$

6. The centripetal force is related to mass m, speed v, and the radius r of the path by the equation

$$F_c = \frac{mv^2}{r}$$

7. Setting the centripetal force equal to the magnetic force allows us to solve for the radius of the path r:

$$r = \frac{mv^2}{F_c}$$
$$= \frac{(1.67 \times 10^{-27} \text{ kg})(6.2 \times 10^6 \text{ m/s})^2}{5.0 \times 10^{-17} \text{ N}}$$
$$= 1.3 \times 10^3 \text{ m}$$

PROBLEM 76   The Lorentz Force

Before we begin...

1. An expression for the net force due to the electric field acting upon the charged particle is

$$\vec{F} = q\vec{E}$$

2. An expression for the net force due to the magnetic field upon the charged particle is

$$F = qvB \sin \theta$$

where θ is the angle between $\vec{v}$ and $\vec{B}$.

Using the right-hand rule, we know that this force is directed perpendicular to both the magnetic field and the velocity.

3. The acceleration $\vec{a}$ is equal to the net force $\vec{F}$ divided by the mass m.

4. The given information is

x component of the $\vec{E}$ field $= 2.5$ V/m

y component of the $\vec{E}$ field $= 5.0$ V/m

z component of the $\vec{B}$ field $= 0.4$ T

$\vec{v} = 10$ m/s

$e = -1.60 \times 10^{-19}$ C

$m_e = 9.11 \times 10^{-31}$ kg

Solving the problem

5. Evaluating the components of the electric force, we find

x direction of $\vec{F}_E = (-1.60 \times 10^{-19} \text{ C})(2.5 \text{ V/m})$
$\qquad = -4.0 \times 10^{-19}$ N

y direction of $\vec{F}_E = (-1.60 \times 10^{-19} \text{ C})(5.0 \text{ V/m})$
$\qquad = -8.0 \times 10^{-19}$ N

Core Concepts in College Physics Workbook

6. Evaluating the magnetic force $\vec{F}_B$,

$$F_B = qvB \sin \theta$$

where θ is the angle between the B field and the velocity of the electron. In this case, θ is 90°, so $\sin \theta = 1$.

$$F_B = (1.60 \times 10^{-19} \text{ C})(10 \text{ m/s})(0.40 \text{ T})$$
$$= 6.4 \times 10^{-19} \text{ N in the } y \text{ direction}$$

The right-hand rule, together with the electron's negative charge, indicates $\vec{F}_B$ points in the $+y$ direction.

7. In the x direction, there is only a contribution from the electric force

$$F_x = -4.0 \times 10^{-19} \text{ N}$$

In the y direction,

$$F_y = F_{Ey} + F_{By}$$
$$= -8.0 \times 10^{-19} \text{ N} + 6.4 \times 10^{-19} \text{ N}$$
$$= -1.6 \times 10^{-19} \text{ N}$$

8. In component form, Newton's second law is

$$F_x = ma_x \qquad F_y = ma_y$$

Solving for the acceleration,

$$a_x = \frac{F_x}{m} \qquad a_y = \frac{F_y}{m}$$

$$a_x = \frac{-4.0 \times 10^{-19} \text{ N}}{9.11 \times 10^{-31} \text{ kg}} = -4.4 \times 10^{11} \text{ m/s}^2$$

$$a_y = \frac{-1.6 \times 10^{-19} \text{ N}}{9.11 \times 10^{-31} \text{ kg}} = -1.8 \times 10^{11} \text{ m/s}^2$$

PROBLEM 77 Ampère's Law

Before we begin...

1. The magnetic force per unit length on a long, straight wire carrying a current $\vec{I}$ in a magnetic field $\vec{B}$ is

$$F/L = IB \sin \theta$$

We know that the wire carries a current $\vec{I} = 2.0$ amps into the page, so we just need to find the magnetic field $\vec{B}$ at point A, in order to calculate the force per unit length.

2.

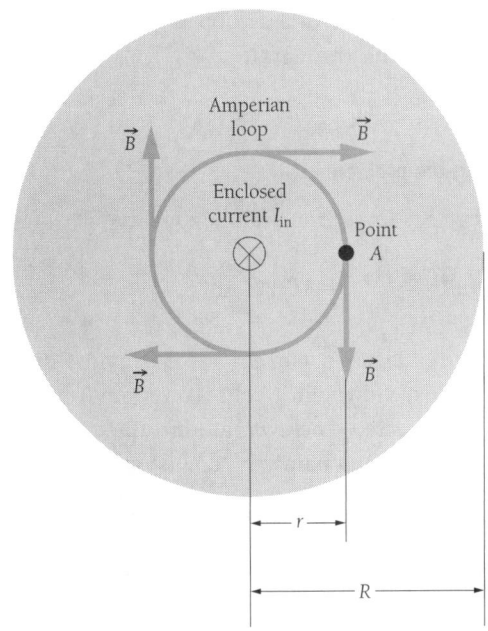

The amperian loop shown in the diagram passes through point A and shares the circular symmetry of the bundle of wires, so $\vec{B}$ should have the same magnitude everywhere around the loop. By the right-hand rule, the direction of $\vec{B}$ should be clockwise around the loop, as shown.

(continued on next page . . .)

3. Not all 100 wires thread the amperian loop in the diagram. The loop is a circle of radius $r = 0.20$ cm, while the bundle is a circle of radius $R = 0.50$ cm. Since the area of a circle is proportional to its radius squared, the loop encloses a fraction of $(r^2/R^2) = 0.16$ of the bundle. Thus 16 of the 100 wires pass through the amperian loop, carrying a current of

$$I_{in} = 16(2.0 \text{ amps}) = 32.0 \text{ amps}$$

4. Since the magnetic field has a constant magnitude everywhere on the amperian loop, and is everywhere tangent to this loop, by Ampère's law, the product of the magnetic field around the amperian loop and the circumference of the loop is μ_0 times the current $\vec{I}_{in}$ passing through it

$$BS = \mu_0 I_{in}$$

Solving the problem

5. Since $S = 2\pi r$, Ampère's law yields

$$B = \frac{\mu_0 I_{in}}{2\pi r} = 3.20 \times 10^{-3} \text{ T}$$

The wire carries a current $I = 2.0$ amps, directed into the page, through a field of strength $B = 3.20 \times 10^{-3}$ T, directed clockwise around the center of the bundle (and thus pointing "down" at point A).

6. The force per unit length on the wire is thus

$$F/L = IB \sin \theta = 6.40 \times 10^{-3} \text{ N/m}$$

directed toward the center of the bundle. This force tends to hold the wires of the bundle together, in other words, although it is probably not strong enough to make much difference if other forces were attempting to pull the bundle apart.

PROBLEM 78 Magnetic Flux and Gauss's Law for Magnetism

Before we begin...

1. The magnetic flux is calculated by the relationship

$$\Phi = BA \cos \theta$$

for a flat surface with a constant magnetic field passing through it.

2.

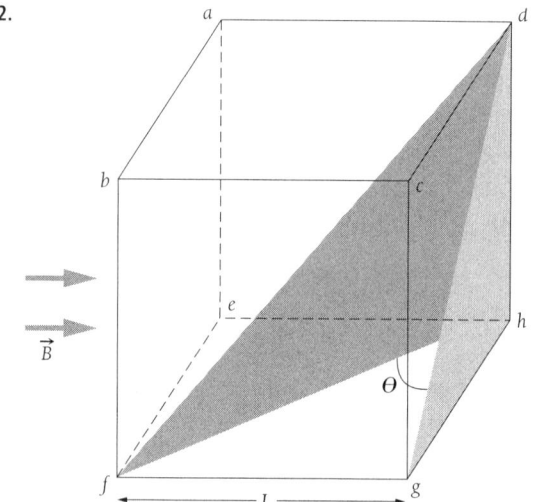

Solving the problem

3. The area of the projection of *dfhd* onto *dcgh* is $\frac{1}{2}L^2$; therefore the magnetic flux is

$$\Phi = BA \cos \theta = B(\tfrac{1}{2}L^2) = \tfrac{1}{2}BL^2$$

4. Because projecting *acfa* onto *abfe* also gives a projected area of $\frac{1}{2} L^2$, the magnetic flux in the second part of the description is also

$$\Phi = BA \cos \theta = B(\tfrac{1}{2}L^2) = \tfrac{1}{2}BL^2$$

PROBLEM 79 Faraday's Law of Induction and Lenz's Law

Before we begin...

1. The given information is

$B = 3.3 \times 10^{-5}$ T $L = 1.0$ m $\omega = 5\pi$ rad/s

Solving the problem

2. Substituting into the equation for the *emf* of a rotating bar and evaluating, we find

$\varepsilon = \tfrac{1}{2} B \omega L^2$
$= (0.5)(3.3 \times 10^{-5}\text{ T})(5\pi \text{ rad/s})(1.0 \text{ m})^2$
$= 2.6 \times 10^{-4}$ V

PROBLEM 80 Faraday's Law

Before we begin...

1. The given information is

$A = 4.0 \times 10^{-4}$ m^2

$r = 2.0$ m

$I_i = 100$ A in the positive y direction

$I_f = 100$ A in the negative y direction

$\Delta t = 0.083$ s

2. The average induced *emf* will equal the average rate of change of the magnetic flux—that is, the change in flux $\Delta \Phi$ divided by the elapsed time Δt. We know Δt, so we must find the initial and final magnetic fluxes through the small loop.

Magnetic flux depends both on the magnetic field and on the area of the loop. We know the area, so we must find the initial and final magnetic fields at the position of the loop.

There is a well-known expression for the magnetic field caused by a steady current in a long, straight wire. Thus, we can work our way backward through all the subproblems we've listed to arrive at the average induced *emf*.

Solving the problem

3. The magnetic field at a distance r from a long, straight wire carrying current I has magnitude

$$B = \mu_0 \frac{I}{2\pi r}$$

and points in a direction which "circulates around" the current, pointing along the fingers of the right hand when the thumb is oriented in the current's direction.

(continued on next page . . .)

Module 12 **The Magnetic Field**

4. In the initial state, with the current flowing in the $+y$ direction, the magnetic field has magnitude

$$B_i = (4\pi \times 10^{-7} \text{ T·m/A}) \frac{(100 \text{ A})}{2\pi(2.0 \text{ m})}$$
$$= 1.0 \times 10^{-5} \text{ T}$$

By the right-hand rule, the direction of $\vec{B}_i$ is into the page. The final magnetic field $\vec{B}_f$ has the same magnitude, but points out of the page since the current creating it has reversed direction.

5. The initial and final magnetic fields are both perpendicular to the loop's surface, so the flux through the loop in both cases has magnitude

$$\Phi_B = BA$$
$$= (1.0 \times 10^{-5} \text{ T})(4.0 \times 10^{-4} \text{ m}^2)$$
$$= 4.0 \times 10^{-9} \text{ T·m}^2$$

The initial flux points into the page, and the final flux points out of the page, so, using the suggested sign convention to indicate direction of flux, we write

$$\Phi_{Bi} = -4.0 \times 10^{-9} \text{ T·m}^2$$
$$\Phi_{Bf} = +4.0 \times 10^{-9} \text{ T·m}^2$$

6. The net change in flux during the time interval is

$$\Delta \Phi_B = \Phi_{Bf} - \Phi_{Bi} = 8.0 \times 10^{-9} \text{ T·m}^2$$

7. By Faraday's law, the average induced *emf* in the loop during the time interval Δt is given by the average rate of change of the magnetic flux through the loop:

$$\langle emf \rangle = \frac{\Delta \Phi_B}{\Delta t}$$
$$= \frac{8.0 \times 10^{-9} \text{ T·m}^2}{0.083 \text{ s}}$$
$$= 9.6 \times 10^{-8} \text{ V}$$

(Yes, a Tesla-meter-squared-per-second equals a volt.)

The problem gives no details of how the current changed from its initial state to its final state. Thus, while we can find the net change in magnetic flux over the time interval, we do not know exactly how quickly the magnetic flux was changing at any particular instant. By Faraday's law, this means that we do not know the exact induced *emf* at any point in time, only the average rate of change over the interval.

PROBLEM 82 Circuit Analysis and Kirchhoff's Laws

Before we begin...

1. For resistors in simple series, the equivalent resistance is

$$R_S = R_1 + R_2 + \ldots$$

2. For resistors in simple parallel, the equivalent resistance is

$$\frac{1}{R_P} = \frac{1}{R_1} + \frac{1}{R_2} + \ldots$$

Solving the problem

3. The resistors R and $5.0\ \Omega$ are connected in series.

4. Their equivalent resistance is $R_1 = R + 5.0\ \Omega$.

5.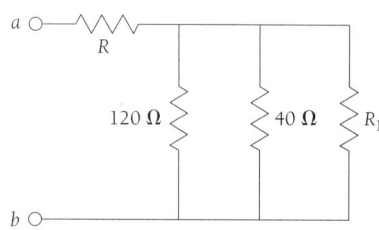

6. In the redrawn diagram, we observe that R_1, the $120\ \Omega$ resistor, and the $40\ \Omega$ resistor are connected in parallel. The equivalent resistance of these elements is found by

$$\frac{1}{R_2} = \frac{1}{R_1} + \frac{1}{40\ \Omega} + \frac{1}{120\ \Omega}$$

7.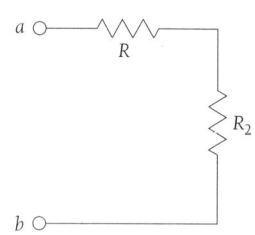

8. In this last diagram, R_2 and the resistor R are connected in series.

$$R_E = R + R_2$$

9. Substituting the values for R_2 and subsequently R_1 into the equation, we find

$$R_E = R + \frac{1}{\frac{1}{R_1} + \frac{1}{40\ \Omega} + \frac{1}{120\ \Omega}}$$

so

$$75\ \Omega = R + \frac{1}{\frac{1}{(R + 5.0\ \Omega)} + \frac{1}{40\ \Omega} + \frac{1}{120\ \Omega}}$$

Simplifying the equation

$$75\ \Omega = R + \frac{1}{\frac{1}{(R + 5.0\ \Omega)} + \frac{1}{30\ \Omega}}$$

becomes

$$75\ \Omega = R + \frac{30\ \Omega(R + 5.0\ \Omega)}{(30\ \Omega + (R + 5.0\ \Omega))}$$

This becomes

$$R^2 - 10R - 2475 = 0$$

When this equation is solved, we find the roots $55\ \Omega$ and $-45\ \Omega$. Because the resistance must be a positive quantity, we have $R = 55\ \Omega$.

PROBLEM 83 Circuit Analysis and Kirchhoff's Laws

Before we begin...

1. The sum of the currents entering any junction must equal the sum of the currents leaving that same junction.

The sum of the potential differences across all of the elements that constitute any closed loop in a circuit is zero.

Solving the problem

2. Applying the junction rule, we have

$$I_2 = I_1 + I_3$$

3. Applying the loop rule to the left loop, we have

$$-(3\ \Omega)I_1 - (5\ \Omega)I_2 + (5.0\ V) = 0$$

For the right loop, we find

$$+(7\ \Omega)I_3 + (5\ \Omega)I_2 - (10.0\ V) = 0$$

4. Solving the equations by the method of substitution yields

$$I_1 = 0.141\ A$$
$$I_2 = 0.915\ A$$
$$I_3 = 0.774\ A$$

5. Evaluating the potential difference between a and b, we set the potential at b equal to zero and travel to a, summing the potential gains and drops as we progress.

$$V_b - V_a = 0 - 10.0\ V - (3.0\ \Omega)(0.141\ A)$$
$$= -10.4\ V$$

If we are going with the current through a resistor, the potential drops. Going against the current through a resistor results in a potential gain.

PROBLEM 84 Capacitors

Before we begin...

1. The equivalent capacitance for capacitors in series is given by

$$\frac{1}{C_S} = \frac{1}{C_1} + \frac{1}{C_2} + \ldots$$

2. The equivalent capacitance for capacitors in parallel is given by

$$C_P = C_1 + C_2 + \ldots$$

3. Capacitors in series have the same charge on each capacitor.

Capacitors in parallel have the same potential difference across each.

4. Capacitance is related to voltage and charge by the equation

$$Q = CV$$

5. The energy stored in a capacitor is computed by

$$U = \frac{1}{2}CV^2 \quad \text{or} \quad U = \frac{1}{2}\left(\frac{Q^2}{C}\right)$$

Solving the problem

6. The 3.0 μF capacitor and the 6.0 μF capacitor are connected in series.

$$\frac{1}{C_2} = \frac{1}{3.0\ \mu F} + \frac{1}{6.0\ \mu F}$$

$$C_2 = 2.0\ \mu F$$

7. The 2.0 μF capacitor and the 4.0 μF capacitor are connected in series

$$\frac{1}{C_3} = \frac{1}{2.0\ \mu F} + \frac{1}{4.0\ \mu F}$$

$$C_3 = 1.33\ \mu F$$

(continued on next page . . .)

8.

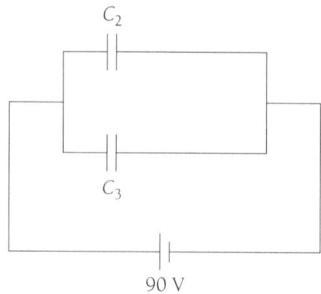

C_2 and C_3 are connected in parallel. The equivalent capacitance C_T is

$$2.0 \ \mu F + 1.3 \ \mu F = 3.3 \ \mu F$$

9. Because C_2 and C_3 are in parallel, they each have the same potential difference, 90 V. The charge on C_2 is

$$q = C_2 V = 180 \ \mu C$$

Because C_2 represents the series of capacitors, 3.0 μF and 6.0 μF, each of these have a charge of 180 μC. The charge on C_3 is

$$q = C_3 V = 120 \ \mu C$$

Because C_3 represents a series of capacitors, 2.0 μF and 4.0 μF, each of these have a charge of 120 μC. By using the relationship

$$V = Q/C$$

we compute the potential differences to be

voltage on 3.0 μF capacitor: $\dfrac{180 \ \mu C}{3.0 \ \mu F} = 60$ V

voltage on 6.0 μF capacitor: $\dfrac{180 \ \mu C}{6.0 \ \mu F} = 30$ V

voltage on 2.0 μF capacitor: $\dfrac{120 \ \mu C}{2.0 \ \mu F} = 60$ V

voltage on 4.0 μF capacitor: $\dfrac{120 \ \mu C}{4.0 \ \mu F} = 30$ V

10. The energy stored by this circuit of capacitors is

$$U = \frac{1}{2} C V^2 = \frac{1}{2} (3.33 \ \mu F)(90 \ V)^2$$

$$= 1.34 \times 10^{-2} \ J$$

PROBLEM 85 Inductors

Before we begin...

1. The potential difference created by an inductor is given by

$$V = L \frac{\Delta I}{\Delta t}$$

2. In the problem description the current has reached its steady-state value; therefore, the rate of change of the current is zero. No potential difference will be measured across the inductor.

3. The energy stored in the magnetic field of an inductor is computed by

$$U = \frac{1}{2} L I^2$$

where L is the inductance and I is the current passing through the inductor.

4. The given information is

$$V = 24.0 \ V \qquad R = 8.00 \ \Omega \qquad L = 4.00 \ H$$

Solving the problem

5. At equilibrium, the entire 24 volt drop occurs across the resistor. Using Ohm's law to compute the value of the equilibrium current, we have

$$I = \frac{V}{R} = \frac{24.0 \ V}{8.00 \ \Omega} = 3.00 \ A$$

6. The energy stored in the magnetic field of the inductor is

$$U = \frac{1}{2} L I^2 = \frac{1}{2} (4.00 \ H)(3.00 \ A)^2 = 18.0 \ J$$

PROBLEM 86 Circuits Containing Resistors, Inductors, and Capacitors

Before we begin...

1. The angular frequency ω_0 is related to the frequency f by

$$\omega_0 = 2\pi f$$

2. The given information is

$f = 120$ Hz $\qquad C = 8.00$ μF

Solving the problem

3. The angular frequency ω_0 is

$$\omega_0 = 2\pi f = (6.28)(120 \text{ Hz}) = 754 \text{ s}^{-1}$$

4. Relating the angular frequency to the inductance and capacitance, we find

$$\omega_0 = \frac{1}{\sqrt{LC}}$$

The value of the inductance is thus

$$L = \frac{1}{(\omega_0^2 C)} = \frac{1}{(754 \text{ s}^{-1})^2 (8.00 \text{ }\mu\text{F})}$$

$$= 0.220 \text{ H}$$

PROBLEM 87 Circuits Containing Resistors, Inductors, and Capacitors

Before we begin...

1. The relationship between ω_0, L, and C is

$$\omega_0 = \frac{1}{\sqrt{LC}}$$

2. The given information is

$L = 2.18$ H $\qquad C = 6.00$ nF

Solving the problem

3. Setting the equations for ω_0 and for β equal to each other and solving for the limiting value of R, we find

$$\frac{R}{2L} = \frac{1}{\sqrt{LC}}$$

such that

$$R = (2)\sqrt{\frac{L}{C}}$$

$$= (2)\sqrt{\frac{(2.18 \text{ H})}{6.00 \text{ nF}}} = 3.81 \times 10^4 \text{ }\Omega$$

Module 14 Geometric Optics

PROBLEM 88 Reflection

Before we begin...

1. The law of reflection states that the angle of incidence equals the angle of reflection as measured from the normal constructed to a reflecting surface.

2.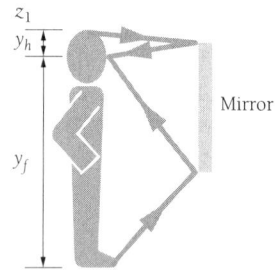

Solving the problem

3. The mirror top must be placed $y_h/2$ or higher for the person still to see the top of his head because the angle of incidence must equal the angle of reflection. (We assume that the top of the head is in the same vertical plane with the eyes.)

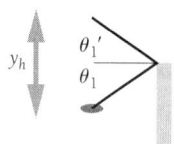

4. The mirror bottom must be placed at least $y_f/2$ below the eyes for the person to see his feet.

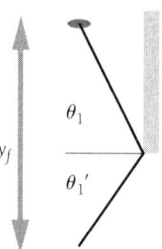

5. The minimum height of the mirror is calculated to be

$$h' = \frac{y_h}{2} + \frac{y_f}{2} = \frac{1}{2}(y_h + y_f) = \frac{1}{2}h$$

PROBLEM 89 Snell's Law

Before we begin...

1.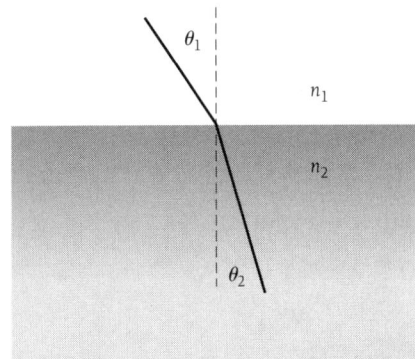

2. Snell's law states that the angle of refraction θ_2 depends on the properties of the two media and on the angle of incidence θ_1 from which the light passes.

$$n_1 \sin \theta_1 = n_2 \sin \theta_2$$

where n represents the index of refraction of the medium.

3. The given information is

$\theta_1 = 30.0°$ $n_1 = 1.00$ $n_2 = 1.33$

Solving the problem

4. Applying Snell's law to solve for the angle at which the light will travel with respect to the normal while in the water, we use

$$n_1 \sin \theta_1 = n_2 \sin \theta_2$$

to evaluate

$$\sin \theta_2 = \frac{n_1 \sin \theta_1}{n_2}$$

$$= \frac{1 \times 0.500}{1.33} = 0.375$$

from which we compute

$$\theta_2 = \sin^{-1}(0.375) = 22.0°$$

PROBLEM 90 Snell's Law

Before we begin...

1. The given information is

$\theta_1 = 45.0°$ $n_1 = 1.00$ $n_2 = 1.52$

Solving the problem

2. Using Snell's law and the given information from above to calculate θ_2, we find

$$n_1 \sin \theta_1 = n_2 \sin \theta_2$$

$$\sin \theta_2 = \frac{n_1 \sin \theta_1}{n_2} = 0.465$$

so

$$\theta_2 = \sin^{-1}(0.465) = 27.7°$$

3. Examining triangle ABC, we observe that

$$\tan \theta_2 = (10 \text{ cm} - x)/(10 \text{ cm})$$

4. Using this relationship to solve for x gives us

$$x = 10 \text{ cm} - (10 \text{ cm}) \tan 27.7° = 4.74 \text{ cm}$$

PROBLEM 91 Total Internal Reflection

Before we begin...

1. The critical angle is the angle of incidence θ_C such that the angle of refraction θ_2 is 90°. The refracted light will move parallel to the boundary. This is possible only when light attempts to move from a medium with a higher index of refraction to a medium with a lower index of refraction.

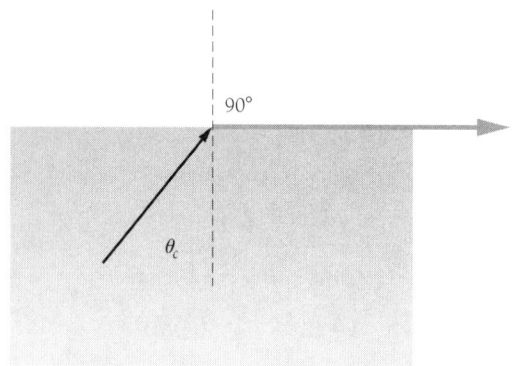

2. The given information is

$n_1 = 1.50$ $n_2 = 1.33$ $\theta_2 = 90°$

Solving the problem

3. Applying Snell's law and the definition of the critical angle to compute θ_C, we find

$$n_1 \sin \theta_C = n_2 \sin 90°$$

Solving for θ_C, we have

$$\sin \theta_C = \frac{n_2(1.00)}{n_1} = \frac{1.33}{1.50} = 0.887$$

from which we compute

$$\theta_C = \sin^{-1}(0.887) = 62.5°$$

PROBLEM 92 Flat and Spherical Mirrors

Before we begin...

1. The lateral magnification is defined as the ratio of the height of the image to the height of the object. For spherical mirrors, it can be computed as

$$M = \frac{-d_i}{d_o}$$

2. The mirror equation is

$$\frac{1}{f} = \frac{1}{d_o} + \frac{1}{d_i}$$

where f is the focal length, d_o the object distance, and d_i the image distance from the mirror.

3. Upright images are virtual. They appear behind the viewing surface of the mirror. The image distance of a virtual image is negative. A concave (or converging) mirror has a positive focal length.

4. The given information is

$$M = +4 \qquad f = +40 \text{ cm}$$

Solving the problem

5. Substituting the relation

$$M = -\frac{d_i}{d_o}$$

into the mirror equation yields

$$\frac{1}{f} = \frac{1}{d_o} + \frac{1}{(-Md_o)} \quad \text{or} \quad d_o = f\left(1 - \frac{1}{M}\right)$$

Substituting the values $f = +40$ cm and $M = +4$, we find

$$d_o = +40\left(1 - \frac{1}{4}\right)$$
$$= 40\left(\frac{3}{4}\right)$$
$$= 30 \text{ cm}$$

PROBLEM 93 Thin Lenses

Before we begin...

1. If an image is projected onto a screen, it must be a real image. Virtual images cannot be projected.

2. The thin lens equation states

$$\frac{1}{f} = \frac{1}{d_o} + \frac{1}{d_i}$$

where f is the focal length, d_o the object distance, and d_i the image distance from the lens.

3. Lateral magnification is defined as the ratio of the height of the image to the height of the object. For thin lenses, it can be computed as

$$M = \frac{-d_i}{d_o}$$

4. The given information is

$$d_o = +32 \text{ cm} \qquad d_i = +8.0 \text{ cm}$$

(Remember that, because the image is real, the image distance is positive.)

Solving the problem

5. Substituting into the thin lens equation, we find

$$\frac{1}{f} = \frac{1}{32 \text{ cm}} + \frac{1}{8.0 \text{ cm}}$$

or

$$f = \frac{+32 \text{ cm}}{5} = +6.4 \text{ cm}$$

Because the focal length is positive, the lens is converging.

6. The lateral magnification is calculated as

$$M = \frac{-d_i}{d_f} = \frac{-8.0 \text{ cm}}{32 \text{ cm}} = -0.25$$

In other words, the image is inverted and one quarter the size of the object.

Module 14 **Geometric Optics**

APPENDIX

Reference Tables

Some Fundamental Constants[a]

Quantity	Symbol	Value[b]
Atomic mass unit	u	1.660 540 2(10) × 10^{-27} kg 931.434 32(28) MeV/c^2
Avogadro's number	N_A	6.022 136 7(36) × 10^{23} (mol)$^{-1}$
Bohr magneton	$\mu_B = \dfrac{e\hbar}{2m_e}$	9.274 015 4(31) × 10^{-24} J/T
Bohr radius	$a_0 = \dfrac{\hbar^2}{m_e e^2 k_e}$	0.529 177 249(24) × 10^{-10} m
Boltzmann's constant	$k_B = R/N_A$	1.380 658(12) × 10^{-23} J/K
Compton wavelength	$\lambda_c = \dfrac{h}{m_e c}$	2.426 310 58(22) × 10^{-12} m
Deuteron mass	m_d	3.343 586 0(20) × 10^{-27} kg 2.013 553 214(24) u
Electron mass	m_e	9.109 389 7(54) × 10^{-31} kg 5.485 799 03(13) × 10^{-4} u 0.510 999 06(15) MeV/c^2
Electron-volt	eV	1.602 177 33(49) × 10^{-19} J
Electron charge	e	1.602 177 33(49) × 10^{-19} C
Gas constant	R	8.314 510(70) J/K · mol
Gravitational constant	G	6.672 59(85) × 10^{-11} N·m^2/kg^2
Hydrogen ground state	$E_0 = \dfrac{m_e e^4 k_e^2}{2\hbar^2} = \dfrac{e^2 k_e}{2a_0}$	13.605 698(40) eV
Josephson frequency-voltage ratio	$2e/h$	4.835 976 7(14) × 10^{14} Hz/V

Some Fundamental Constants[a] (continued)

Quantity	Symbol	Value[b]
Magnetic flux quantum	$\Phi_0 = \dfrac{h}{2e}$	$2.067\ 834\ 61(61) \times 10^{-15}$ Wb
Neutron mass	m_n	$1.674\ 928\ 6(10) \times 10^{-27}$ kg $1.008\ 664\ 904(14)$ u $939.565\ 63(28)$ MeV/c^2
Nuclear magneton	$\mu_n = \dfrac{e\hbar}{2m_p}$	$5.050\ 786\ 6(17) \times 10^{-27}$ J/T
Permeability of free space	μ_0	$4\pi \times 10^{-7}$ N/A^2 (exact)
Permittivity of free space	$\epsilon_0 = 1/\mu_0 c^2$	$8.854\ 187\ 817 \times 10^{-12}$ C^2/N·m^2 (exact)
Planck's constant	h $\hbar = h/2\pi$	$6.626\ 075(40) \times 10^{-34}$ J·s $1.054\ 572\ 66(63) \times 10^{-34}$ J·s
Proton mass	m_p	$1.672\ 623(10) \times 10^{-27}$ kg $1.007\ 276\ 470(12)$ u $938.272\ 3(28)$ MeV/c^2
Quantized Hall resistance	h/e^2	$25812.805\ 6(12)\ \Omega$
Rydberg constant	R_H	$1.097\ 373\ 153\ 4(13) \times 10^7$ m^{-1}
Speed of light in vacuum	c	$2.997\ 924\ 58 \times 10^8$ m/s (exact)

[a]These constants are the values recommended in 1986 by CODATA, based on a least-squares adjustment of data from different measurements. For a more complete list, see Cohen, E. Richard, and Barry N. Taylor, *Rev. Mod. Phys.* 59:1121, 1987

[b]The numbers in parentheses for the values below represent the uncertainties in the last decimal places.

Module 13 Electric Circuits

PROBLEM 81 Voltage, Resistance, and Ohm's Law

Before we begin...

1. The power delivered to electric circuits is computed by using

$$P = IV$$

where I represents the electric current and V is the potential difference.

2. The given information is

$V = 12.0$ volts cost/kW·h = $0.06

rating = 55.0 A·h

Solving the problem

3. Expressing power as current times potential difference, the expression for the energy delivered in terms of current, voltage, and time t is written as

$$U = Pt = IVt$$

4. The dimensions of the equation are

[current] × [potential difference] × [time]

5. We have to multiply the rating of the battery [current] × [time] by [potential difference].

6. Rearranging the expression for energy, we determine that

$$U = IVt = (It)V$$
$$= (55.0 \text{ A·h})(12.0 \text{ V}) = 660 \text{ W·h}$$

7. To calculate the cost of this energy, we convert the W·h to kW·h by

$$660 \text{ W·h} = (660 \text{ W·h})(1 \text{ kW}/1000 \text{ W})$$
$$= 0.660 \text{ kW·h}$$

The cost of the total energy is computed to be

$$\text{cost} = (0.660 \text{ kW·h})(\$0.06/\text{kW·h})$$
$$= \$0.0396$$

The cost of the energy is certainly not what dictated the cost of the battery!

Solar System Data

Body	Mass (kg)	Mean Radius (m)	Period (s)	Distance from Sun (m)
Mercury	3.18×10^{23}	2.43×10^{6}	7.60×10^{6}	5.79×10^{10}
Venus	4.88×10^{24}	6.06×10^{6}	1.94×10^{7}	1.08×10^{11}
Earth	5.98×10^{24}	6.37×10^{6}	3.156×10^{7}	1.496×10^{11}
Mars	6.42×10^{23}	3.37×10^{6}	5.94×10^{7}	2.28×10^{11}
Jupiter	1.90×10^{27}	6.99×10^{7}	3.74×10^{8}	7.78×10^{11}
Saturn	5.68×10^{26}	5.85×10^{7}	9.35×10^{8}	1.43×10^{12}
Uranus	8.68×10^{25}	2.33×10^{7}	2.64×10^{9}	2.87×10^{12}
Neptune	1.03×10^{26}	2.21×10^{7}	5.22×10^{9}	4.50×10^{12}
Pluto	$\approx 1.4 \times 10^{22}$	$\approx 1.5 \times 10^{6}$	7.82×10^{9}	5.91×10^{12}
Moon	7.36×10^{22}	1.74×10^{6}	—	—
Sun	1.991×10^{30}	6.96×10^{8}	—	—

Physical Data Often Used[a]

Average Earth-Moon distance	3.84×10^{8} m
Average Earth-Sun distance	1.496×10^{11} m
Average radius of the Earth	6.37×10^{6} m
Density of air (20°C and 1 atm)	1.29 kg/m^3
Density of water (20°C and 1 atm)	1.00×10^{3} kg/m^3
Free-fall acceleration	9.80 m/s^2
Mass of the Earth	5.98×10^{24} kg
Mass of the Moon	7.36×10^{22} kg
Mass of the Sun	1.99×10^{30} kg
Standard atmospheric pressure	1.013×10^{5} Pa

[a]These are the values of the constants as used in the text.

Some Prefixes for Powers of Ten

Power	Prefix	Abbreviation
10^{-18}	atto	a
10^{-15}	femto	f
10^{-12}	pico	p
10^{-9}	nano	n
10^{-6}	micro	μ
10^{-3}	milli	m
10^{-2}	centi	c
10^{-1}	deci	d
10^{1}	deka	da
10^{2}	hecto	h
10^{3}	kilo	k
10^{6}	mega	M
10^{9}	giga	G
10^{12}	tera	T
10^{15}	peta	P
10^{18}	exa	E

Standard Abbreviations and Symbols of Units

Abbreviation	Unit
A	ampere
Å	angstrom
u	atomic mass unit
atm	atmosphere
Btu	British thermal unit
C	coulomb
°C	degree Celsius
cal	calorie
deg	degree (angle)
eV	electron volt
°F	degree Fahrenheit
F	farad
ft	foot
G	gauss
g	gram
H	henry
h	hour
hp	horsepower
Hz	hertz
in.	inch
J	joule
K	kelvin
kcal	kilocalorie
kg	kilogram
kmol	kilomole
lb	pound
m	meter
min	minute
N	newton
Pa	pascal
rev	revolution

(*continued on next page . . .*)

Standard Abbreviations and Symbols of Units
(continued)

Abbreviation	Unit
s	second
T	tesla
V	volt
W	watt
Wb	weber
μm	micrometer
Ω	ohm

Mathematical Symbols Used in the Text and Their Meaning

Symbol	Meaning
$=$	is equal to
$\equiv$	is defined as
$\neq$	is not equal to
$\propto$	is proportional to
$>$	is greater than
$<$	is less than
$\gg$ ($\ll$)	is much greater (less) than
$\approx$	is approximately equal to
Δx	the change in x
$\sum_{i=1}^{N} x_i$	the sum of all quantities x_i from $i = 1$ to $i = N$
$|x|$	the magnitude of x (always a positive quantity)
$\Delta x \rightarrow 0$	Δx approaches zero
$\dfrac{dx}{dt}$	the derivative of x with respect to t
$\dfrac{\partial x}{\partial t}$	the partial derivative of x with respect to t
$\int$	integral

Reference Tables

Useful Conversions

Length
12 in. = 1 ft

3 ft = 1 yd

1 yd = 0.9144 m

1 Å = 10^{-10} m

1 μm = 1 μ = 10^{-6} m = 10^4 Å

1 lightyear = 9.461×10^{15} m

Area
1 m^2 = 10^4 cm^2 = 10.76 ft^2

1 ft^2 = 0.0929 m^2 = 144 $in.^2$

1 $in.^2$ = 6.452 cm^2

Volume
1 m^3 = 10^6 cm^3 = 6.102×10^4 $in.^3$

1 ft^3 = 1728 $in.^3$ = 2.83×10^{-2} m^3

1 liter = 1000 cm^3 = 1.0576 qt = 0.0353 ft^3

1 ft^3 = 7.481 gal = 28.32 liters = 2.832×10^{-2} m^3

1 gal = 3.786 liters = 231 $in.^3$

Mass
1000 kg = 1 t (metric ton)

Velocity
1 mi/min = 60 mi/h = 88 ft/s

Acceleration
1 m/s^2 = 3.28 ft/s^2 = 100 cm/s^2

1 ft/s^2 = 0.3048 m/s^2 = 30.48 cm/s^2

Pressure
1 bar = 10^5 N/m^2 = 14.50 $lb/in.^2$

Energy
931.5 MeV/c^2 is equivalent to 1 u

Power
1 hp = 550 ft · lb/s = 0.746 kW

1 W = 1 J/s = 0.738 ft · lb/s

1 Btu/h = 0.293 W

The Greek Alphabet

Name	Uppercase	Lowercase
Alpha	A	α
Beta	B	β
Gamma	Γ	γ
Delta	Δ	δ
Epsilon	E	ϵ
Zeta	Z	ζ
Eta	H	η
Theta	Θ	θ
Iota	I	ι
Kappa	K	κ
Lambda	Λ	λ
Mu	M	μ
Nu	N	ν
Xi	Ξ	ξ
Omicron	O	o
Pi	Π	π
Rho	P	ρ
Sigma	Σ	σ
Tau	T	τ
Upsilon	Y	υ
Phi	Φ	ϕ
Chi	X	χ
Psi	Ψ	ψ
Omega	Ω	ω

Conversion Factors

LENGTH

	m	cm	km	in.	ft	mi
1 meter	1	10^2	10^{-3}	39.37	3.281	6.214×10^{-4}
1 centimeter	10^{-2}	1	10^{-5}	0.3937	3.281×10^{-2}	6.214×10^{-6}
1 kilometer	10^3	10^5	1	3.937×10^4	3.281×10^3	0.6214
1 inch	2.540×10^{-2}	2.540	2.540×10^{-5}	1	8.333×10^{-2}	1.578×10^{-5}
1 foot	0.3048	30.48	3.048×10^{-4}	12	1	1.894×10^{-4}
1 mile	1609	1.609×10^5	1.609	6.336×10^4	5280	1

MASS

	kg	g	slug	u
1 kilogram	1	10^3	6.852×10^{-2}	6.024×10^{26}
1 gram	10^{-3}	1	6.852×10^{-5}	6.024×10^{23}
1 slug	14.59	1.459×10^4	1	8.789×10^{27}
1 atomic mass unit	1.660×10^{-27}	1.660×10^{-24}	1.137×10^{-28}	1

TIME

	s	min	h	day	year
1 second	1	1.667×10^{-2}	2.778×10^{-4}	1.157×10^{-5}	3.169×10^{-8}
1 minute	60	1	1.667×10^{-2}	6.994×10^{-4}	1.901×10^{-6}
1 hour	3600	60	1	4.167×10^{-2}	1.141×10^{-4}
1 day	8.640×10^4	1440	24	1	2.738×10^{-3}
1 year	3.156×10^7	5.259×10^5	8.766×10^3	365.2	1

	SPEED			
	m/s	cm/s	ft/s	mi/h
1 meter/second	1	10^2	3.281	2.237
1 centimeter/second	10^{-2}	1	3.281×10^{-2}	2.237×10^{-2}
1 foot/second	0.3048	30.48	1	0.6818
1 mile/hour	0.4470	44.70	1.467	1

Note: 1 mi/min = 60 mi/h = 88 ft/s.

	FORCE		
	N	dyn	lb
1 newton	1	10^5	0.2248
1 dyne	10^{-5}	1	2.248×10^{-6}
1 pound	4.448	4.448×10^5	1

	WORK, ENERGY, HEAT		
	J	erg	ft · lb
1 joule	1	10^7	0.7376
1 erg	10^{-7}	1	7.376×10^{-8}
1 ft · lb	1.356	1.356×10^7	1
1 eV	1.602×10^{-19}	1.602×10^{-12}	1.182×10^{-19}
1 cal	4.186	4.186×10^7	3.087
1 Btu	1.055×10^3	1.055×10^{10}	7.779×10^2
1 kWh	3.600×10^6	3.600×10^{13}	2.655×10^6
	eV	cal	Btu	kWh
1 joule	6.242×10^{18}	0.2389	9.481×10^{-4}	2.778×10^{-7}
1 erg	6.242×10^{11}	2.389×10^{-8}	9.481×10^{-11}	2.778×10^{-14}
1 ft · lb	8.464×10^{18}	0.3239	1.285×10^{-3}	3.766×10^{-7}
1 eV	1	3.827×10^{-20}	1.519×10^{-22}	4.450×10^{-26}
1 cal	2.613×10^{19}	1	3.968×10^{-3}	1.163×10^{-6}
1 Btu	6.585×10^{21}	2.520×10^2	1	2.930×10^{-4}
1 kWh	2.247×10^{25}	8.601×10^5	3.413×10^2	1

	PRESSURE		
	Pa	dyn/cm²	atm
1 pascal	1	10	9.869×10^{-6}
1 dyne/centimeter²	10^{-1}	1	9.869×10^{-7}
1 atmosphere	1.013×10^5	1.013×10^6	1
1 centimeter mercury*	1.333×10^3	1.333×10^4	1.316×10^{-2}
1 pound/inch²	6.895×10^3	6.895×10^4	6.805×10^{-2}
1 pound/foot²	47.88	4.788×10^2	4.725×10^{-4}
	cm Hg	lb/in.²	lb/ft²
1 newton/meter²	7.501×10^{-4}	1.450×10^{-4}	2.089×10^{-2}
1 dyne/centimeter²	7.501×10^{-5}	1.450×10^{-5}	2.089×10^{-3}
1 atmosphere	76	14.70	2.116×10^3
1 centimeter mercury*	1	0.1943	27.85
1 pound/inch²	5.171	1	144
1 pound/foot²	3.591×10^{-2}	6.944×10^{-3}	1

*At 0°C and at a location where the acceleration due to gravity has its "standard" value, 9.80665 m/s².

Symbols, Dimensions, and Units of Physical Quantities

Quantity	Common Symbol	Unit*	Dimensions†	Unit in Terms of Base SI Units
Acceleration	a	m/s²	L/T²	m/s²
Amount of substance	n	mole		mol
Angle	θ, ϕ	radian (rad)	1	
Angular acceleration	$\vec{\alpha}$	rad/s²	T^{-2}	s^{-2}
Angular frequency	ω	rad/s	T^{-1}	s^{-1}
Angular momentum	L	kg·m²/s	ML²/T	kg·m²/s
Angular velocity	$\vec{\omega}$	rad/s	T^{-1}	s^{-1}
Area	A	m²	L²	m²
Atomic number	Z			
Capacitance	C	farad (F)(= Q/V)	Q²T²/ML²	A²·s⁴/kg·m²
Charge	q, Q, e	coulomb (C)	Q	A·s

Symbols, Dimensions, and Units of Physical Quantities (continued)

Quantity	Common Symbol	Unit*	Dimensions†	Unit in Terms of Base SI Units
Charge density				
Line	λ	C/m	Q/L	A·s/m
Surface	σ	C/m^2	Q/L^2	A·s/m^2
Volume	ρ	C/m^3	Q/L^3	A·s/m^3
Conductivity	σ	1/Ω·m	Q^{2}T/ML^3	A^2·s^3/kg·m^3
Current	I	AMPERE	Q/T	A
Current density	$\vec{J}$	A/m^2	Q/T^2	A/m^2
Density	ρ	kg/m^3	M/L^3	kg/m^3
Dielectric constant	κ			
Displacement	s	METER	L	m
Distance	d, h			
Length	ℓ, L			
Electric dipole moment	$\vec{p}$	C·m	QL	A·s·m
Electric field	$\vec{E}$	V/m	ML/QT^2	kg·m/A·s^3
Electric flux	Φ	V·m	ML^3/QT^2	kg·m^3/A·s^3
Electromotive force	$\vec{\varepsilon}$	volt (V)	ML^2/QT^2	kg·m^2/A·s^3
Energy	E, U, K	joule (J)	ML^2/T^2	kg·m^2/s^2
Entropy	S	J/K	ML^2/T^2·K	kg·m^2/s^2·K
Force	$\vec{F}$	newton (N)	ML/T^2	kg·m/s^2
Frequency	f, v	hertz (Hz)	T^{-1}	s^{-1}
Heat	Q	joule (J)	ML^2/T^2	kg·m^2/s^2
Inductance	L	henry (H)	ML^2/Q^2	kg·m^2/A^2·s^2
Magnetic dipole moment	$\vec{\mu}$	N·m/T	QL^2/T	A·m^2
Magnetic field	$\vec{B}$	tesla (T) (=Wb/m^2)	M/QT	kg/A·s^2
Magnetic flux	Φ_m	weber (Wb)	ML^2/QT	kg·m^2/A·s^2
Mass	m, M	KILOGRAM	M	kg
Molar specific heat	C	J/mol·K		kg·m^2/s^2·mol·K
Moment of inertia	I	kg·m^2	ML^2	kg·m^2
Momentum	$\vec{p}$	kg·m/s	ML/T	kg·m/s
Period	T	s	T	s
Permeability of space	μ_0	N/A^2 (=H/m)	ML/Q^{2}T	kg·m/A^2·s^2
Permittivity of space	ϵ_0	C^2/N·m^2 (=F/m)	Q^{2}T^2/ML^3	A^2·s^4/kg·m^3

Symbols, Dimensions, and Units of Physical Quantities (continued)

Quantity	Common Symbol	Unit*	Dimensions†	Unit in Terms of Base SI Units
Potential (voltage)	V	volt (V)(=J/C)	ML^2/QT^2	kg·m²/A·s³
Power	P	watt (W)(=J/s)	ML^2/T^3	kg·m²/s³
Pressure	P, p	pascal (Pa) = (N/m²)	M/LT^2	kg/m·s²
Resistance	R	ohm (Ω)(=V/A)	ML^2/Q^2T	kg·m²/A²·s³
Specific heat	c	J/kg·K	$L^2/T^2·K$	m²/s²·K
Temperature	T	KELVIN	K	K
Time	t	SECOND	T	s
Torque	$\vec{\tau}$	N·m	ML^2/T^2	kg·m²/s²
Speed	v	m/s	L/T	m/s
Volume	V	m³	L^3	m³
Wavelength	λ	m	L	m
Work	W	joule (J)(=N·m)	ML^2/T^2	kg·m²/s²

*The base SI units are given in uppercase letters.
†The symbols M, L, T, and Q denote mass, length, time, and charge, respectively.

SI Base Units

Base Quantity	SI BASE UNIT Name	Symbol
Length	meter	m
Mass	kilogram	kg
Time	second	s
Electric current	ampere	A
Temperature	kelvin	K
Amount of substance	mole	mol
Luminous intensity	candela	cd

Some Derived SI Units

Quantity	Name	Symbol	Expression in Terms of Base Units	Expression in Terms of Other SI Units
Plane angle	radian	rad	m/m	
Frequency	hertz	Hz	s^{-1}	
Force	newton	N	$kg \cdot m/s^2$	J/m
Pressure	pascal	Pa	$kg/m \cdot s^2$	N/m^2
Energy: work	joule	J	$kg \cdot m^2/s^2$	$N \cdot m$
Power	watt	W	$kg \cdot m^2/s^3$	J/s
Electric charge	coulomb	C	$A \cdot s$	
Electric potential (emf)	volt	V	$kg \cdot m^2/A \cdot s^3$	W/A
Capacitance	farad	F	$A^2 \cdot s^4/kg \cdot m^2$	C/V
Electric resistance	ohm	Ω	$kg \cdot m^2/A^2 \cdot s^3$	V/A
Magnetic flux	weber	Wb	$kg \cdot m^2/A \cdot s^2$	$V \cdot s$
Magnetic field intensity	tesla	T	$kg/A \cdot s^2$	Wb/m^2
Inductance	henry	H	$kg \cdot m^2/A^2 \cdot s^2$	Wb/A